OXFORD MODERN LANGUAGES
AND LITERATURE MONOGRAPHS

Technique and Technology

Script, Print, and Poetics in France, 1470–1550

ADRIAN ARMSTRONG

CLARENDON PRESS · OXFORD

This book has been printed digitally and produced in a standard specification in order to ensure its continuing availability

OXFORD
UNIVERSITY PRESS

Great Clarendon Street, Oxford OX2 6DP

Oxford University Press is a department of the University of Oxford.
It furthers the University's objective of excellence in research, scholarship,
and education by publishing worldwide in

Oxford New York

Auckland Cape Town Dar es Salaam Hong Kong Karachi
Kuala Lumpur Madrid Melbourne Mexico City Nairobi
New Delhi Shanghai Taipei Toronto
With offices in
Argentina Austria Brazil Chile Czech Republic France Greece
Guatemala Hungary Italy Japan South Korea Poland Portugal
Singapore Switzerland Thailand Turkey Ukraine Vietnam

Published in the United States
by Oxford University Press Inc., New York

Reprinted 2007

ISBN 978-0-19-815989-6

Cover illustration: Jean Lemaire de Belges visualizing the scenes he is about to narrate in the *Couronne Margaritique*: ÖNB, Cod. 3441, fo. 2ᵛ (photo: Bildarchiv/ÖNB Wien)

For Rachel

PREFACE

This book contributes to two fields in which much excellent research has appeared in recent decades: that of the *Grand Rhétoriqueur* poets of fifteenth- and sixteenth-century France; and that of the relationship between literature and its physical presentation. The focus is on the poetry of three major *rhétoriqueurs*, whose careers together embrace the period of transmission from manuscript to print in France; I examine the changing ways in which the self-conscious aspects of these poets' writing are manifested on the page, foregrounded or otherwise by layout, illustration, paratext, and other elements of book form. My study should thus, I hope, be of interest not solely to specialists in *rhétoriqueur* poetry, but also to bibliographers and codicologists; to scholars with interests in the phenomenon of literary self-consciousness; and to all those concerned with the literary implications of the materiality of texts.

In a study concerned with the collaborative processes of book production, and the guiding role of paratext, it is both apposite and pleasurable to devote some of my own paratext to thanking those whose collaboration, in diverse respects, has been essential to the production of this book. For their valuable comments and suggestions, I am extremely grateful to my examiners Richard Cooper and Pierre Jodogne; to the OUP readers Mike Freeman and Ian Maclean; and to those who responded to my research at conferences and seminars. Help and advice, both scholarly and practical, was generously provided by many individuals, among whom my debt is greatest to Jacques Abélard, Jennifer Britnell, Cynthia Brown, David Cowling, Jean Devaux, Philip Ford, Simon Gaunt, Roy and Jan Harbour, Sarah Kay, Ann Moss, Roger Pensom, Anne Schoysman, the late Dudley Wilson, and Michel Zink.

It has been a delight to explore the resources of numerous libraries; I owe especial gratitude to the British Library; the Bibliothèque Nationale de France; the Bibliothèque Royale Albert Ier, Brussels; the Bodleian Library and Taylor Institution, Oxford; the University Library and Gonville and Caius College Library, Cambridge; the John Rylands University Library, Manchester; the Musée Condé, Chantilly; the Médiathèque François-Mitterrand,

Poitiers; and the Bibliothèque de l'Arsenal and Institut de Recherche et d'Histoire des Textes, Paris. My research would have been impossible without the awards of a Major State Studentship from the British Academy, and of a Research Fellowship at Trinity Hall, Cambridge, with whose Master and Fellows I spent a particularly happy year in 1994–5. Publication was facilitated by grants from the Research and Graduate School in Arts, and the Research and Graduate Support Unit, of the University of Manchester. I am grateful to Oxford University Press for their kind permission to reuse material from a previous article (noted in the Bibliography), 'The Practice of Textual Transmission: Jean Molinet's *Ressource du petit peuple*'.

The interest and encouragement of my friends Catherine Constable, Kevin Sullivan, and David Willis, and of my colleagues at Manchester, especially David Adams, Martin Crowley, John Flynn, and Henry Phillips, have been of inestimable value. Finally, my heartfelt thanks go to the supervisor of my D.Phil. thesis, Jane Taylor, for her intellectual rigour, unfailing encouragement, and great personal generosity; to my parents Ray and Mary, and my brother Matthew, for their constant moral and occasional practical support; and to Rachel Jones, my sharpest critic and closest friend.

A.A.

CONTENTS

LIST OF PLATES

ABBREVIATIONS

BFLUS	*Bulletin de la Faculté des Lettres de l'Université de Strasbourg*
BHR	*Bibliothèque d'humanisme et Renaissance*
BL	British Library, London
BM	Bibliothèque Municipale
BNF	Bibliothèque Nationale de France, Paris
BR	Bibliothèque Royale Albert Ier, Brussels
BSG	Bibliothèque Sainte-Geneviève, Paris
CAIEF	*Cahiers de l'Association Internationale des Études Françaises*
FD	*Les Faictz et dictz de Jean Molinet*, ed. N. Dupire (Paris, 1936–9)
IRHT	Institut de Recherche et d'Histoire des Textes, Paris
JMRS	*Journal of Medieval and Renaissance Studies*
MA	*Le Moyen Âge*
MF	*Le Moyen Français*
MFM	Médiathèque François-Mitterrand, Poitiers
NRSS	*Nouvelle Revue du seizième siècle*
Œ	J. Lemaire de Belges, *Œuvres*, ed. J. Stecher (Louvain, 1882–91)
ÖNB	Österreichische Nationalbibliothek, Vienna
RN	*Romance Notes*

INTRODUCTION

The interpreters like to think they're the aristocrats of this institution, and affect to look down on us selectors. But as you must know, this is pure vanity on their part. Anyone with the least gumption can see that without us here in Selection, Interpretation would be like a mill without any wheat. We're the ones who supply them with all their raw material. We are the basis of their success.[1]

So speaks the head of the Selection department in *The Palace of Dreams*, Ismail Kadare's haunting vision of the police state. He is addressing Mark-Alem, a young nobleman beginning his career in the Tabir Sarrail, the eponymous Palace, where dreams are collected from every corner of the Empire, selected, classified, and interpreted, with a view to finding the Master-Dreams which foretell the Empire's future. But the words of Mark-Alem's superior also describe the institution of literary studies. Critics (interpreters) depend for their texts upon editors, textual critics, bibliographers; in short, the Selection department. Naturally, this division is artificial: as Mark-Alem himself soon realizes, the work of selection itself involves interpretation. Yet, for the sake of convenience, the departments generally remain separate, to the detriment of literary scholarship. While it may be difficult to integrate the Palace's various tasks, the issue is vital and urgent for a discipline faced with both methodological overspecialization and a computer-driven media revolution with profound implications for the study and practice of literature.[2] It is in the context of an earlier media revolution that the present study contributes to the fusion of the Palace's departments. I address the interaction of literary forms with technologies of textual production in a context eminently suitable for this approach; that of the poets of fifteenth- and sixteenth-century France commonly known as the *Grands Rhétoriqueurs*, and of the evolution of book

[1] I. Kadare, *The Palace of Dreams*, tr. J. Vrioni, tr. B. Bray (London, 1993), 24.

[2] R. Pensom, 'Zumthor and After: A Survey of Some Current Trends in the Reading of Old French Literature', discusses methodological specialization in medieval studies. On information technology and literature, see G. P. Landow, *Hypertext: The Convergence of Contemporary Critical Theory and Technology* (Baltimore, 1992).

production from a manuscript-based to a print-based process in the same milieu.[3]

A proper appreciation of *rhétoriqueur* poetic practice reveals the extent to which it was intimately related to the physical form of the book, whether in manuscript or print. Whatever the ideological orientation and stylistic predilections of individual poets,[4] the *rhétoriqueurs* tended to valorize technical difficulty and the overcoming of formal constraints, manipulating versification and language in often dazzling feats of virtuosity. Early scholars decried such techniques as facile gimmickry.[5] More recently, Paul Zumthor has suggested that the *rhétoriqueurs*, whose freedom to compose was limited by their professional dependence upon patrons, engaged in linguistic and prosodic experimentation in order to undermine inherited poetic language, and thereby to 'repersonnaliser le rapport de l'écrivant à l'écriture'.[6] Neither of these two approaches, however, adequately accounts for the implications of form in *rhétoriqueur* poetry. The *rhétoriqueurs* evince a striking awareness of their vocation as moralists, historiographers, and propagandists;[7] it is in these capacities that they produced their most significant and influential work. Far from subverting their language, formal virtuosity enhances it, as several scholars have amply shown.[8] It is most judicious to see the *rhétoriqueurs'* use of form and language as reflecting two key aspects of their aesthetic. First, the significance of their works' subject-matter required a corresponding degree of formal elaboration to do it justice. Second, contemporary notions of poetry made formal complexity a criterion of professional excellence: the more difficult the

[3] The term *Grand Rhétoriqueur* is erroneous: see P. Jodogne, 'Les "Rhétoriqueurs" et l'humanisme: Problème d'histoire littéraire', 150–8. Nevertheless, it has become consecrated by scholarly usage, and is universally understood as designating the poets and practices on which I focus.

[4] On the contrasts between French and Burgundian *rhétoriqueurs*, see C. Thiry, 'Rhétoriqueurs de Bourgogne, rhétoriqueurs de France: Convergences, divergences?'

[5] See, for example, H. Guy, *Histoire de la poésie française au XVIe siècle*, i: *L'École des rhétoriqueurs* (Paris, 1910).

[6] P. Zumthor, *Le Masque et la lumière: La Poétique des grands rhétoriqueurs* (Paris, 1978), 42–55 (54).

[7] See Jodogne, 'Les "Rhétoriqueurs"', 161–5.

[8] See C. Thiry, 'La Poétique des grands rhétoriqueurs'; F. Cornilliat, *'Or ne mens': Couleurs de l'éloge et du blâme chez les 'Grands Rhétoriqueurs'* (Paris, 1994), 27–36; J. Devaux, *Jean Molinet, indiciaire bourguignon* (Paris, 1996), 9–18.

prosodic exercise, the more talented the poet who could accomplish it.[9]

One symptom of this attitude to form is an exploitation of the page as visual space: many *rhétoriqueur* texts incorporate acrostics, rebuses, or stanzas which could be read in several directions.[10] Such techniques appear not only in ludic verse, but also in political and religious pieces; they indicate that the *rhétoriqueurs* maximized their works' rhetorical sophistication by working on the visual dimension of language as well as its other aspects. One can thus hardly do justice to *rhétoriqueur* poetry without paying close attention to its presentation on the page in the original manuscripts and printed books. Modern critical editions, despite their obvious necessity and utility, unfortunately blur this vital issue, since considerations of reader-friendliness and economy require layouts to be normalized. Noël Dupire, for instance, in his edition of Jean Molinet's poetry, rationalizes the manuscript layout of a complex rebus-poem, *Ung present fait a l'empereur*, into a rigid typographical grid which poses far less of an interpretative challenge to the reader.[11] The ossification imposed by critical editing is always regrettable, if unavoidable; for the *rhétoriqueurs*, though, it is catastrophic, disabling proper analysis. Returning to the original books, then, is a step as essential as it is overdue.

Critical approaches of this kind have recently found considerable favour in literary studies. In diverse historical periods, the materiality of texts has come to constitute a legitimate object of analysis.[12] Editorial practice has been

[9] Pierre Fabri reflects this view when discussing equivocal rhymes: see P. Fabri, *Le Grand et vrai art de pleine rhétorique*, ed. A. Héron (Rouen, 1890), ii. 17–23. Fabri values the greatest possible contrast between rhyme-words' phonetic similarity and semantic difference. If rhyme-words are spelled differently, or if monosyllables and polysyllables are rhymed together, this contrast is reduced, and the poet's achievement less considerable. See Cornilliat, '*Or ne mens*', 190–1.

[10] Examples are cited in C.-G. Dubois, 'L'Invention littéraire et les jeux de langage: Jeux de nombres, jeux de sons, jeux de sens'; D. Hüe, 'Enjeux et emblèmes de la poésie à lectures multiples'; G. Gros, *Le Poète marial et l'art graphique: Étude sur les jeux de lettres dans les poèmes pieux du Moyen Âge* (Caen, 1993).

[11] *FD*, i. 269. Further references to this edition will be incorporated into the text. See Ch. 1 below for discussion of this manuscript layout.

[12] See, for example, J. J. McGann, *The Textual Condition* (Princeton, 1991); M. Couturier, *Textual Communication: A Print-Based Theory of the Novel* (London, 1991); and R. Chartier, *L'Ordre des livres: Lecteurs, auteurs, bibliothèques en Europe entre XIVe et XVIIIe siècle* (Aix-en-Provence, 1992).

re-evaluated,[13] while medievalists increasingly acknowledge the significance of manuscript transmission and scribal practices for literary interpretation and critical editing.[14] In the case of the *rhétoriqueurs*, it is all the more valuable to address texts' physical manifestations in books because these poets' activity coincided with the shift from manuscript to print. This shift had fundamental results not only for books' dissemination and presentation, but also for the practice of writing itself.

Printing's broad cultural implications have been widely studied.[15] Contemporaries quickly recognized the new invention's utility in rapidly producing vast numbers of identical copies, at a price consequently lower than that of manuscripts; equally, they acknowledged the dangers resulting from the swift distribution of faulty or heretical texts.[16] Compared to scribal copying, printing mediated more heavily between a text and its readers, since a greater variety of textual producers was involved;[17] colophons were prominent in early printed books, underlining the channels through which texts now passed on their way to their audience. Marks and colophons also advertised printers' products; techniques of publicity quickly developed, to permit books to enjoy the high sales upon which printers depended for financial survival.[18] In addition, the visual appearance of books changed. The earliest printed books resembled manuscripts in many respects: decoration and punctuation were not provided by printers, but

[13] See, for example, J. J. McGann, *A Critique of Modern Textual Criticism* (Chicago, 1983); B. Cerquiglini, *Éloge de la variante: Histoire critique de la philologie* (Paris, 1989); D. Hult, 'Reading It Right: The Ideology of Text Editing'.

[14] See, for example, E. Kennedy, 'The Scribe as Editor'; S. Huot, *From Song to Book: The Poetics of Writing in Old French Lyric and Lyrical Narrative Poetry* (Ithaca, NY, 1987), and *The 'Romance of the Rose' and its Medieval Readers: Interpretation, Reception, Manuscript Transmission* (Cambridge, 1993); S. G. Nichols, 'The New Philology: Introduction: Philology in a Manuscript Culture'; S. L. Hindman (ed.), *Printing the Written Word: The Social History of Books, circa 1450–1520* (Ithaca, NY, 1991).

[15] See, for example, L. Febvre and H.-J. Martin, *L'Apparition du livre* (Paris, 1958); R. Hirsch, *Printing, Selling and Reading 1450–1550* (Wiesbaden, 1967); E. Eisenstein, *The Printing Press as an Agent of Change: Communications and Cultural Transformations in Early-Modern Europe* (Cambridge, 1979); W. J. Ong, *Orality and Literacy: The Technologizing of the Word* (London, 1982), ch. 5; H.-J. Martin and R. Chartier (eds.), *Histoire de l'édition française*, i: *Le Livre conquérant* (Paris, 1982); Hindman, *Printing*.

[16] H. Widmann, *Vom Nutzen und Nachteil der Erfindung des Buchdrucks—aus der Sicht der Zeitgenossen des Erfinders* (Mainz, 1973); Eisenstein, *Printing Press*, i. 44–51, 71–80.

[17] Ong, *Orality and Literacy*, 122; A. Parent, *Les Métiers du livre à Paris au XVIe siècle (1535–1560)* (Geneva, 1974), 120.

[18] Eisenstein, *Printing Press*, i. 57–60.

added by hand.[19] However, the standardization of copies, and increased regularity of presentation, had significant effects. Referencing aids could be employed much more effectively than in manuscripts; punctuation was standardized and refined; pictures could be repeated precisely, to the great benefit of science.[20] Over the course of the sixteenth century, printers came to take more decisions regarding punctuation and decoration, which it had originally been the reader's or owner's responsibility to execute.[21] Perhaps the most striking visual contrast between manuscripts and printed books lay in the emergence of the title-page, a development intimately related to printing. Whereas a manuscript's owner had typically commissioned the volume for himself, and therefore knew its contents, printers needed to attract a wide public, which was often unfamiliar with the texts they published. Consequently, title-pages were developed as a useful means of both identifying and advertising texts.[22]

Besides affecting books' distribution and appearance, printing also promoted changes in literary composition, as authors sought to come to terms with a new readership. Whereas the public in manuscript culture comprised a relatively cohesive and familiar group of confrères or patrons, the public opened up by printing was unfamiliar, much larger and more diverse, and no longer necessarily shared an author's literary, cultural, or political assumptions. To avoid misreadings on the part of this new public, many writers composed explanatory prefaces to guide readers, or commented explicitly on their discourse to clarify their viewpoint. Such developments have received a measure of scholarly attention.[23] However, the most detailed and sophisticated treatment of printing's literary implications deals specifically with the

[19] C. F. Bühler, *The Fifteenth-Century Book: The Scribes, the Printers, the Decorators* (Philadelphia, 1960), ch. 1; P. Saenger and M. Heinlen, 'Incunable Description and its Implication for the Analysis of Fifteenth-Century Reading Habits'.

[20] Ong, *Orality and Literacy*, 123–8; Eisenstein, *Printing Press*, i. 51–5, 88–107; M. B. Parkes, *Pause and Effect: An Introduction to the History of Punctuation in the West* (Aldershot, 1992), 87–92; W. M. Ivins, *Prints and Visual Communication* (London, 1953), 1–50.

[21] Saenger and Heinlen, 'Incunable Description', 252–5.

[22] R. Hirsch, 'Title Pages in French Incunables, 1486–1500'.

[23] See, for example, B. Lydgate, 'Printing, Narrative, and the Genesis of the Rabelaisian Novel', 355–6; 'Mortgaging One's Work to the World: Publication and the Structure of Montaigne's *Essais*', 213–14; W. Kennedy, 'Petrarchan Audiences and Print Technology', 15–18; D. N. Losse, *Sampling the Book: Renaissance Prologues and the French 'Conteurs'* (Lewisburg, Pa., 1994), 22, 77.

rhétoriqueurs, amply illustrating the importance of such approaches for an understanding of these poets' work. In a series of articles and a major recent book, *Poets, Patrons, and Printers*, Cynthia Brown has convincingly demonstrated that printing altered the *rhétoriqueurs'* perception of their role.[24] In manuscript culture, the *rhétoriqueurs* were financially dependent on patrons, and hence presented themselves in their work in self-effacing terms, as essentially subordinate figures. As printing became more widespread, it changed the relationships between poets, patrons, and book producers. *Rhétoriqueurs* developed proprietorial attitudes towards their writing, as manifested in their occasional applications for privileges (legal documents temporarily restricting the right to publish a text to named individuals, usually printers). In paratextual features such as title-pages, colophons, and illustrations, authors assumed greater prominence;[25] a similar tendency characterizes later *rhétoriqueurs'* use of first-person narration, in which the intratextual narrative voice appears increasingly to coincide with the historical author.[26]

Brown's welcome and stimulating research both influences and complements my own. As will become clear, my study corroborates Brown's findings in *Poets, Patrons, and Printers*, often by analysing texts other than those she addresses. However, our two studies differ considerably in focus and scope. Brown concentrates primarily upon printing's socio-cultural implications for authors who were actively involved in publishing their material, in a period extending approximately from the 1490s to 1520. While I deal to an extent with the same questions, my approach also embraces two further aspects of book form. First, I examine the importance of books as indices of the reception of *rhétoriqueur*

[24] C. J. Brown, 'The Rise of Literary Consciousness in Late Medieval France: Jean Lemaire de Belges and the *Rhétoriqueur* Tradition'; 'Du manuscrit à l'imprimé en France: Le Cas des Grands Rhétoriqueurs'; 'L'Éveil d'une nouvelle conscience littéraire en France à la grande époque de transition technique: Jean Molinet et son moulin poétique'; 'Text, Image, and Authorial Self-Consciousness in Late Medieval Paris'; *Poets, Patrons, and Printers: Crisis of Authority in Late Medieval France* (Ithaca, NY, 1995).

[25] For 'paratext', see G. Genette, *Seuils* (Paris, 1987). The term is problematic, since it embraces both material accompanying a text and metatextual discourse transmitted separately ('epitext'). Moreover, Genette's assumption that paratext is sanctioned by the author (14) is inappropriate to 15th- and 16th-century conditions of textual production. Nevertheless, the term remains useful in designating material annexed to a text.

[26] For all these issues, see especially the relevant chapters in *Poets, Patrons, and Printers*, which develops the observations made in Brown's various articles.

poetry; second, I discuss the ways in which texts' material presentation influences the reader's perception of their formal properties. In order to treat these issues adequately, I consider fewer writers than Brown, but extend my observations over a longer period, as I explain below.

My study's various concerns are interrelated, in that they reflect a central preoccupation. By tracing *rhétoriqueur* poetry's physical manifestations in manuscript and print, I demonstrate that the evolution of book production contributes greatly to a change in the mode of literary self-consciousness which the poets evince. I understand 'self-consciousness' to denote the various ways in which a text may foreground its author's creative investment in it. This definition permits the recognition that, as recent studies of self-consciousness in the contemporary novel have shown, texts may proclaim their 'literariness' in diverse ways.[27] They may highlight not only their processes of elaboration or their authors' identity, but also their own linguistic surface, either by explicitly thematizing it or by using striking language which draws readers' attention to 'the semantic, syntactic, and often also phonetic texture of the words which actually serve to structure as well as constitute the work'.[28]

Several scholars have discussed particular manifestations of self-consciousness in *rhétoriqueur* poetry, which are each embraced by my definition. Zumthor stresses the ways in which the poets draw attention to the linguistic medium; Brown sees self-consciousness in *rhétoriqueur* writing largely in terms of poets' 'increasing use of self-promotional strategies'.[29] Various studies note the *rhétoriqueurs'* tendency to praise and commemorate fellow-poets, to thematize their position as writers, or to construct first-person narrators which appear at least partly to reflect the historical author, thus reinforcing the poet's presence in his work.[30]

[27] See, for example, L. Hutcheon, *Narcissistic Narrative: The Metafictional Paradox* (London, 1984), ch. 1; P. Waugh, *Metafiction: The Theory and Practice of Self-Conscious Fiction* (London, 1984), 1–61; B. McHale, *Postmodernist Fiction* (London, 1987). V. Erlich, *Russian Formalism: History–Doctrine* (The Hague, 1955), chs. 10–11, outlines 'literariness' in Russian Formalism.

[28] Hutcheon, *Narcissistic Narrative*, 119.

[29] Brown, *Poets, Patrons, and Printers*, 6.

[30] M.-R. Jung, 'Simon Gréban, rhétoriqueur', notes these tendencies in one of the earliest *rhétoriqueurs*. See also M. B. Winn, 'In Pursuit of the *Acteur* in *La Chasse et le départ d'Amours*'; M.-R. Jung, 'Les *Douze dames de rhétorique*'; J.-C. Muhlethaler, *Poétiques du quinzième siècle: Situation de François Villon et Michault Taillevent* (Paris, 1983), 11–16; J. Cerquiglini,

Other self-conscious techniques are also apparent in *rhétoriqueur* poetry. Linguistic virtuosity is manifested in the use of polysemous allegory as well as in prosodic patterning. A text may comment on its own elaboration, whether explicitly, in metadiscursive fashion, or implicitly; it may contain prominent intertextual allusions relating it to the genre off which it plays, and thereby underlining its literariness. Poets may exploit books' physical form to various ends. The use of expressive layout, or of other presentational features which stress a text's materiality, highlights the literary work's artificiality, and hence also the poet's compositional activity. Elements of textual apparatus such as marginalia reinforce the reader's impression that the writer of a historical or polemical piece has adequately researched his subject, and can discuss it with proper authority. Flattering liminary pieces, or the production of a sumptuous presentation manuscript, enable poets to further their material interests and highlight the work in which they have engaged on behalf of their current or prospective patron. The explicit attribution of a text to its author, in manuscript or print, implies at least the interest of book producers in the poet's identity, and may betoken authorial self-promotion; moreover, the author is clearly much more prominent in an attributed text than in an anonymous piece, so that readers obtain a far stronger impression of the writer's presence behind the work.

What is common to these different techniques, all of which I discuss in this study, is that they stress the activity of writing, and render the text 'less the mirror of nature, more an *artifact*, visibly a *made* thing'.[31] To a greater or lesser extent, they are present in all literature; to recognize their importance in *rhétoriqueur* poetry does not imply that the latter somehow resembles the postmodern novel. Rather, to acknowledge that all these techniques are aspects of literary self-consciousness provides a coherent conceptual framework within which to assess how *rhétoriqueur* poetry develops as technologies of book production change. I trace the different ways in which self-consciousness is manifested in the work of successive poets, in the different media.

'L'Écriture louche: La Voie oblique chez les Grands Rhétoriqueurs'; Y. LeBlanc, *Va lettre va: The French Verse Epistle (1400–1550)* (Birmingham, Ala., 1995), ch. 3; D. Cowling, *Building the Text: Architecture as Metaphor in Late Medieval and Early Modern France* (Oxford, 1998), ch. 5.

[31] McHale, *Postmodernist Fiction*, 30 (original emphasis).

Broadly, as printed books supplant manuscripts, self-consciousness is exhibited less through formalistic techniques, and more through 'personal' or metadiscursive preoccupations. Technical and socio-cultural aspects of printing promote this development. In order to make their writing clearer to the new, unfamiliar public, authors explain and comment on their arguments, and draw attention to significant features of their work in liminary material. Printing also enables and indeed encourages authors to signal their identity on title-pages, thus asserting their authorship and authority over the text more decisively than is possible in manuscripts. Conversely, *rhétoriqueur* formalism is not well conveyed in print: compositors have less freedom to organize a page than scribes, and are dependent on a printer's stock of type and woodcuts. Consequently, rebus-poems and the like are more difficult to present without some degree of visual simplification. All these factors help explain why later *rhétoriqueurs* like Jean Bouchet and Pierre Gringore, both of whom were involved in the publication of their works, display less prosodic virtuosity and more direct techniques of self-promotion than earlier, more manuscript-bound poets like Molinet.

My study focuses on selected works by three poets: Molinet, Jean Lemaire de Belges, and Bouchet. These are the ideal figures on which to base my analysis, for two reasons. First, their work best displays the great variety of self-conscious techniques I have outlined; they were all prolific poets, whose work covers a wide range of different forms and functions. Second, their successive literary lifetimes, which the dates in my study's title indicate approximately, not only combine to embrace almost the whole timespan of the *rhétoriqueur* school, but also correspond to three distinct phases in the evolution of book production, and of authors' involvement in it. Molinet was active between 1464 and his death in 1507, based largely in the Burgundian Netherlands.[32] His output is perhaps the most formally sophisticated of any *rhétoriqueur*'s, and contains a high proportion of poems which exploit visual presentation. Molinet's works were transmitted largely in manuscript, but a significant number of printed editions also appeared; most of these were posthumous, but Molinet may have been involved in certain texts' publication. Lemaire, whose known

[32] For Molinet's life, see N. Dupire, *Jean Molinet: La Vie–les œuvres* (Paris, 1932); Devaux, *Jean Molinet*, 115–51.

literary activity spans the period 1498–1515, worked for various patrons.[33] He composed both rhetorically intricate pieces, sometimes employing expressive layouts, and works which display reflexive or apparently autobiographical features. Though his output depended far more than Molinet's on printing, over which he apparently had some influence in several cases, many of his works also survive in manuscript form. I devote two chapters each to Molinet and Lemaire, one each for manuscripts and printed books. Bouchet's career spanned some sixty years until his death in the late 1550s.[34] In his work, formal virtuosity is secondary to the employment of sophisticated allegory, and to the construction of a prominent first-person narratorial voice, often apparently representing Bouchet himself. Never a court poet, Bouchet worked overwhelmingly in the printed medium from his home town of Poitiers, and participated in the publication of most of his works long after their first editions. Since too few manuscripts of his poetry exist to merit discussion separately from the editions, I consider Bouchet's work in a single chapter.

Each of these poets wrote in both prose and verse. However, verse is intrinsically more deeply implicated than prose in questions of layout and textual production. Layout plays a decisive role in readers' perception and interpretation of verse: lineation, and the division into stanzas, organize language in ways which cannot but be significant.[35] Moreover, the visual presentation of verse requires book producers to make more choices than does that of prose. How much spacing, for instance, should be left between stanzas? Should refrains be abbreviated after their first occurrence, or shorter lines be indented? Such considerations mean that scribes and printers can exert more influence upon readers' perception of verse than is possible in the case of prose. Accordingly, the texts I discuss are in either verse or *prosimetrum*. Prose texts are only considered in cases where their presentation influences that of poetry: this applies particularly to editions of Lemaire's works, which frequently transmit poetry alongside prose pieces, including his major historical work, *Les Illustrations de Gaule et singularités*

[33] For Lemaire's career, see P. Jodogne, *Jean Lemaire de Belges, écrivain franco-bourguignon* (Brussels, 1972), ch. 3.

[34] J. Britnell, *Jean Bouchet* (Edinburgh, 1986), 1–5.

[35] For verse layout's importance and evolution, see J. Culler, *Structuralist Poetics: Structuralism, Linguistics and the Study of Literature* (London, 1975), 183; Parkes, *Pause and Effect*, ch. 8; J. Lennard, *The Poetry Handbook: A Guide to Reading Poetry for Pleasure and Practical Criticism* (Oxford, 1996), chs. 3, 5.

de Troie. Two principles govern the selection of material, which I explain in each chapter. The texts analysed illustrate, first, the chronological range and varied receptions, in manuscript and print, of each author's poetry; second, the diverse self-conscious techniques which these writers deploy.

My analysis embraces all the factors which may contribute to texts' presentation: layout, illustration, and other aspects of what Roger Chartier has termed *mise en livre*, a book's visual organization;[36] paratext, such as title-pages or prefaces; and volumes' circumstances of origin (date, place, scribe or printer). Where authorial influence on textual production seems likely, I treat volumes as aspects of the poet's literary project; otherwise, they are regarded as indices of the ways in which both readers and book producers received the texts they transmit. An accurate assessment of these issues must obviously take account of contemporary practice and expectations in two areas, that of literature and that of book production. I shall outline my approach to both these areas in turn.

Clearly, it is impossible fully to reconstruct the horizon of expectations of the *rhétoriqueurs'* contemporaries; any attempt to recuperate a past culture's norms must recognize the epistemological problems of such an exercise.[37] Nevertheless, reconstruction of this kind is a practical necessity. Consequently, when I refer to the reader's experience of a text, illustration, or layout, I employ the notion of an 'ideal reader'. This concept, formulated by Jonathan Culler, is 'a theoretical construct, perhaps best thought of as a representation of the central notion of acceptability'; it represents not 'the immediate and spontaneous reactions of individual readers but the meanings which they are willing to accept as both plausible and justified when they are explained'.[38] I base criteria of acceptability primarily upon existing knowledge of fifteenth- and sixteenth-century literary

[36] R. Chartier, 'Du livre au lire', 102–3.

[37] R. C. Holub, *Reception Theory: A Critical Introduction* (London, 1984), 60–1.

[38] Culler, *Structuralist Poetics*, 124. When using pronouns to refer to 'the reader', I employ the male pronoun throughout, for the sake of convenience; this gendering is unsatisfactory, but probably reflects the *rhétoriqueurs'* readership. Ownership of books of *rhétoriqueur* poetry was predominantly male, according to the limited available evidence. While ownership cannot be equated with the extent of the reading public, it constitutes a significant index to the circulation of texts. See A. H. Schutz, *Vernacular Books in Parisian Private Libraries of the 16th Century According to the Notarial Inventories* (Chapel Hill, NC, 1955); A. Labarre, *Le Livre dans la vie amiénoise du seizième siècle: L'Enseignement des inventaires après décès, 1503–1576* (Paris, 1971).

practice, especially the model of *rhétoriqueur* poetics previously outlined.[39] Where possible, to minimize the dangers of circularity inherent in reconstructing a horizon of expectations from norms which can only be perceived from a twentieth-century perspective, I support the 'ideal reader' construct with factual evidence such as readers' annotations in texts (another benefit of studying contemporary documents).

My own interpretative practices, however, naturally draw upon more recent approaches to literature. The interdisciplinary nature of my study entails the *bricolage* of a new critical vocabulary with which to analyse unexpected conjunctions of literary, bibliographical, and iconographical features, for terms often do not yet exist to describe and assess these conjunctions' effects. Concepts must be borrowed from other disciplines; words must be used figuratively. The result is a liberation from the constraints of the customary lexis, the standard interpretative moves, of literary study; the conscious blurring of terminology facilitates a more penetrating examination of the key issues. For the same reasons, I avoid adopting a single critical perspective, which would risk overlooking significant factors in the interests of a homogenizing method. A deconstructionist analysis, for instance, might focus on how book form can partly subvert *rhétoriqueur* poetry's political messages by drawing attention to linguistic virtuosity, therefore implying that political discourse's values are not self-evident, and drawing attention to the ways in which words drift along associative paths so that language itself is the texts' ultimate referent. However, this fails to recognize that *rhétoriqueurs* accept the *dérives* of wordplay as part of political discourse; politics and poetic self-assertion are not mutually exclusive for them. An eclectic approach is much better suited to the subject, since analysis must begin with the books themselves: each book raises different questions, which lend themselves to examination in whichever terms seem most useful as heuristic tools. Detailed case studies, rather than an all-embracing theory, are the only way in which to address a field whose boundaries—fortunately—cannot be precisely defined.

Several important studies have outlined the norms of *mise en*

[39] Important surveys of the literature of this and neighbouring periods are D. Poirion, *Le Poète et le prince: L'Évolution du lyrisme courtois de Guillaume de Machaut à Charles d'Orléans* (Paris, 1965); Zumthor, *Masque*; Muhlethaler, *Poétiques*; Cornilliat, '*Or ne mens*'.

livre in both manuscript and print: it is with respect to these norms that I assess the physical presentation of the texts I discuss.[40] However, one aspect of my approach to *mise en livre* requires initial clarification. The typography and illustration of early printed books were often determined in large part by the limitations of a printer's stock, and his house style. This tendency is most apparent in the frequent use of stock images, rather than specially designed woodcuts, to illustrate books. Consequently, my examination of printed texts' visual presentation often entails comparing the volume concerned with the printer's other publications, or studying its woodcuts' provenance. Yet this is by no means always the case. Contemporary readers' expectations regarding *mise en livre* are much more likely to have been general, embracing a definite range of possibilities for a text's presentation, than to have been formed on the basis of individual printers' styles. Specific house styles may be significant in certain cases, but for the most part are less relevant to the construction of a readerly horizon of expectations than are the general tendencies evident in early printing.

An aspect of *mise en livre* which merits particular examination is textual illustration, which lends itself especially well to interdisciplinary approaches, and has attracted stimulating research by structuralist critics and their successors.[41] While the links between text and image cannot be considered in isolation from a text's overall physical presentation, the analysis of illustrations requires its own methodology, which must be explained at the outset. As with the study of *mise en livre* in general, terms and methods for interpreting images are not fixed, especially not when these images are linked to literary texts. A process of dynamic *bricolage* must again be followed to describe illustrations' effects adequately. Theories of visual semiotics which attempt to account for all aspects of images by using linguistic models are not only methodologically flawed,[42] but also too reductive to account for images' relationships to accompanying texts. Indeed, general theories

[40] See Martin and Chartier, *Le Livre conquérant*; Parkes, *Pause and Effect*; J. Lemaire, *Introduction à la codicologie* (Louvain-la-Neuve, 1989); Bühler, *The Fifteenth-Century Book*; H.-J. Martin and J. Vezin (eds.), *Mise en page et mise en texte du livre manuscrit* (Paris, 1990). P. Saenger, *Space between Words: The Origins of Silent Reading* (Stanford, Calif., 1997), ch. 15, usefully discusses broader aspects of manuscript culture in the late Middle Ages.

[41] See, for example, R. Barthes, 'Rhétorique de l'image', and the journal *Word and Image*.

[42] See Groupe μ, *Traité du signe visuel* (Paris, 1992), 10–11.

of pictorial resemblance are of little practical use: the manner in which iconic signs represent their referents is not functionally relevant to the reader of an illustrated book. More immediately useful is an awareness of the intericonic context within which images are produced; Ernst Gombrich and others have recognized the importance of familiar images in the creation and interpretation of unfamiliar ones.[43] Accordingly, studies of contemporary iconography constitute an essential basis for my readings.[44]

The choice and content of individual illustrations poses its own problems, which I address through concepts borrowed from various sources. The processes of 'condensation' and 'displacement' outlined in Freud's theory of 'dream-work' provide a useful model for investigating how textual features may be translated into visual images. Disparate motifs or episodes in a text may be combined into a single image, in a process of condensation; a secondary motif in a text may assume unexpected importance in an illustration, by virtue of displacement.[45] From reception theory I adopt the notion of *Unbestimmtheitsstellen*,[46] to refer to elements which may not be described in a text but which cannot be omitted in illustrations, such as a protagonist's physical appearance. It is often valuable to examine how images fill these *Unbestimmtheitsstellen*: they may follow latent hints in the text (for example, by making physical attributes correspond to moral ones), adopt existing pictorial conventions, or, in printed books, simply reuse stock woodcuts. The latter technique poses interpretative problems, as cuts' relevance to the text often seems aleatory. Unless images depict familiar scenes like the Crucifixion, context must determine what they represent. Woodcuts may acquire new captions or be slightly reworked, while cuts repeated within a single book gain a generic significance; this is often true of battle

[43] E. Gombrich, *Art and Illusion: A Study in the Psychology of Pictorial Representation* (5th edn. Oxford, 1977), 72.

[44] Such studies include M. Schapiro, *Words and Pictures: On the Literal and the Symbolic in the Illustration of a Text* (The Hague, 1973); F. Garnier, *Le Langage de l'image au Moyen Âge* (Paris, 1982–8); J. Wirth, *L'Image médiévale: Naissance et développements (VIe–XVe siècle)* (Paris, 1989); D. Russell, *Emblematic Structures in Renaissance French Culture* (Toronto, 1995); G. de Tervarent, *Attributs et symboles dans l'art profane: Dictionnaire d'un langage perdu (1450–1600)* (2nd edn. Geneva, 1997).

[45] Cf. S. Freud, *The Interpretation of Dreams* (London, 1954), v. 506–8. The processes of 'dream-work' are used to elucidate illustrations of the *Roman de Mélusine* in F. Clier-Colombani, *La Fée Mélusine au Moyen Âge: Images, mythes et symboles* (Paris, 1991), 26.

[46] See R. Warning, *Rezeptionsästhetik* (Munich, 1975), 11.

scenes, for instance. In some cases, reused cuts seem to be irrelevant to the text, and to play a purely decorative or structural role.[47] While all these factors must be considered when studying stock images, in specific cases a cut's relevance is often best assessed through the concept of redundancy, which originates in information theory.[48] By establishing which textual themes or motifs are iterated in the cut, it is possible to determine how a reused image may foreground certain textual elements, whether or not the printer was aware of it.

The Tabir Sarrail is a selective institution, separating dreams of different types and different degrees of interest. Throughout my study, I demonstrate that this procedure, though valuable, must be followed by the reintegration of what has been sundered if literary studies are to give a proper account of their object. Texts cannot be assessed without examining their original transmission and presentation; cheap paper manuscripts deserve no less attention than luxury volumes; methodologies must be combined, not employed singly. In Kadare's novel, Mark-Alem eventually completes a major piece of research, 'a long report on the sleep and dreams of the whole Empire';[49] but he comes to realize that to search for experience beyond the Palace must cause his downfall. I hope that, in combining disciplines, in synthesizing approaches, in addressing a subject with implications for the literature of other periods (not least that of the information superhighway), my own research has made the Palace's walls more permeable.

Conventions of quotation and transcription

When considering the version of a text in a particular manuscript or edition, I provide a diplomatic transcription; otherwise, I cite from the relevant critical edition if one exists. Quasi-facsimile reproduction of original texts is inevitably a partial and illusory

[47] See R. S. Luborsky, 'Connections and Disconnections between Images and Texts: The Case of Secular Tudor Book Illustration'; D. Sansy, 'Texte et image dans les incunables français'.

[48] S. R. Suleiman, 'Introduction: Varieties of Audience-Oriented Criticism', 16, explains this notion's relevance to literary studies.

[49] Kadare, *Palace of Dreams*, 187.

exercise, which cannot fully reproduce books' visual properties.[50] Nevertheless, it is an essential means of conveying to readers not only the evolution of orthography and punctuation over the period studied, but also the 'ipseity' of each individual volume. In quotations from original books, the slash / represents a *virgula*; | represents a line break. Abbreviations are retained as far as possible; thus ⁊ denotes the tironian *et*. Red ink is represented by bold type. Irregular signatures in printed books are represented according to standard conventions.[51]

[50] Cerquiglini, *Éloge*, 43.
[51] See P. Gaskell, *A New Introduction to Bibliography* (Oxford, 1972), 330–1.

I

MANUSCRIPTS OF MOLINET'S POETRY

Introduction

During his long career, spent primarily as *indiciaire* of the dukes of Burgundy while based in Valenciennes,[1] Molinet composed a wide variety of poems, with differing forms and functions. In his capacity as official historiographer, he produced encomiastic and occasional poems, including weighty *prosimetrum* pieces on political subjects. However, his prolific output also includes didactic, devotional, parodic, and bawdy verse, and epistles in both verse and prose.[2] It is only to be expected that such a heterogeneous *œuvre* should display a varying range of self-conscious techniques.

The most prominent such technique, and the basis of Molinet's reputation, consists in the virtuoso ways in which poems' language and versification are elaborated. Texts abound in neologisms, loan-words, and punning; verse adopts an impressive variety of stanzaic forms, with rhyme-schemes of often bewildering complexity.[3] Formal devices encompass the whole range of *rhétoriqueur* poetics: some poems alternate Latin with French text; some consist of fragmented lines which can be read in various directions or permutations; some contain acrostics, or pictorial rhymes.[4] Such texts invite a view of their author as an expert craftsman; this image of Molinet's activity is reinforced by his *Art de rhétorique*, which systematizes his concepts of poetic excellence and presents rhetoric as essentially the manipulation of formal patterns.[5]

[1] Dupire, *Jean Molinet*, ch. 1. Devaux, *Jean Molinet*, 25–46, outlines the *indiciaire*'s tasks.

[2] Gros, *Poète marial*, analyses several of Molinet's devotional poems. Comic and bawdy material is discussed in G. Angeli, 'Il "rude engin" di Molinet', 32–44, and L. W. Johnson, *Poets as Players: Theme and Variation in Late Medieval French Poetry* (Stanford, Calif., 1990), ch. 5.

[3] Dupire, *Jean Molinet*, chs. 8–10, catalogues elements of Molinet's language and style.

[4] For examples, see *FD*, i. 269; ii. 812–14, 878. Gros, *Poète marial*, 66–75, 93–4, examines acrostics in Molinet's devotional verse.

[5] On the *Art*, see Dupire, *Jean Molinet*, 61–70; *Recueil d'arts de seconde rhétorique*, ed. E. Langlois (Paris, 1902), pp. lvii–lxxiii; É. Méchoulan, 'Les Arts de rhétorique du XVe siècle: La Théorie, masque de la *theoria*?' The *Art* is edited in *Recueil*, ed. Langlois, 214–52; further references will be to this edition, referred to as *Art*.

Virtuosity of this kind appears in all types of verse, whether political, pious, or scatological.

Other self-conscious elements are relatively uncommon in Molinet's work. Metadiscourse, such as dedications or allusions to the writing process, is usually brief and conventional. It is often confined to poems' final stanzas or sections, in several of which a first-person narrator refers to writing down the allegorical dream or vision he has experienced.[6] Prefatory material appears rarely in Molinet's poetry, in contrast with his lengthy prose works, which include significant prologues;[7] the poems are probably too short to be accompanied by discrete liminary pieces of this kind, as opposed to brief dedications and the like within the text proper. The manuscript transmission of much of the poetry may also partly explain its relative lack of paratext: the familiarity of Molinet's court-based audience perhaps rendered presentational and explanatory discourse otiose, while any material not formally integrated to the text risked being omitted by selective copyists.

The first-person voice in much of Molinet's poetry is generalized, rather than a representation of Molinet as historical author. Apparently autobiographical elements usually operate metaphorically, rather than referentially. Molinet's references to the loss of an eye form part of a late-medieval poetic tradition which associated imperfect eyesight and other physical disabilities with the poet's social position, as a cleric subordinate to noble patrons.[8] Similarly, in the *Complainte sur la mort madame d'Ostrisse* (*FD*, i. 162–80), an introductory account of the first-person narrator's visits to various European courts is probably not a summary of Molinet's own career, but a review of contemporary Europe which praises the court of Mary of Burgundy by comparing it favourably to others. In numerous poems, Molinet employs a

[6] See, for example, *FD*, i. 126, vv. 17–20; 161, vv. 56–9; 180, vv. 473–80.

[7] Molinet's *Chroniques* have two prologues; the second, probably a revised version of the first, appears in fewer manuscripts. See Molinet, *Chroniques*, ed. G. Doutrepont and O. Jodogne (Brussels, 1935–7), i. 25–8; ii. 589–95. The earlier prologue praises the House of Burgundy, while the second outlines Molinet's task as historian; Devaux, *Jean Molinet*, 47–51, examines the prologues' content. Similarly, the *Roman de la Rose moralisé* includes a prologue in which Molinet reflects on love, and on the undertaking of rewriting the romance in prose: *Roman de la Rose moralisé*, Paris, Anthoine Vérard, n.d. (BL, C. 22. c. 2), fos. b1^r–3^r. The *Art de rhétorique* contains a preface in two manuscripts, BNF, mss. fr. 2159 and 2375: see *Art*, 214–15.

[8] Cerquiglini, 'Écriture louche'.

signature which puns on the literal sense of his name (*moulinet*, 'windmill'). Yet even this signature should probably be seen not primarily as a reference to the historical poet, but as part of Molinet's work on language, which, as often in this period, motivates and semantizes even the apparently arbitrary proper name.[9]

Some of Molinet's later poems do present an apparently more individualized, autobiographical authorial voice. Explicit references to Molinet's name gradually replace the punning signature.[10] Moreover, the poet's activity is thematized to a greater extent: whereas the first-person narrator in earlier occasional and didactic pieces was primarily a passive witness of events,[11] in some later works he is a self-conscious figure who alludes to the writing process. These developments are strikingly illustrated in two poems from 1496, the *Gaiges retrenchiés* and *Voiage de Napples* (*FD*, ii. 768–71; i. 277–83). In the *Gaiges*, the poet complains about the halving of his pension to Baudouin de Lannoy, captain of the Château of Lille, whom Molinet hopes will intercede on his behalf with his patron, Philip the Handsome. Through his punning signature, Molinet elaborates the image of a windmill unable to produce poetry for want of grain. He alludes to his activity as poet, and specifically, in a very rare instance of self-quotation,[12] to his earlier critique of war, the *Temple de Mars*:

> Il [sc. ung molinet] a mollut, tout net jusque a l'estrain,
> De Mars le train qui gens d'armes alume,
> En lettres d'or, d'azur, d'argent, d'arain;
> Tant le derrain que le premier gros grain
> Noble et purain a mys a son volume. (769, vv. 17–21)

However, by referring to a windmill in the third person, rather than to himself in the first, Molinet presents his predicament somewhat indirectly.[13] In the *Voiage*, Molinet employs a rare metadiscursive prologue; here, he states the poem's historical

[9] See J. R. Scheidegger, 'La Lettre du nom: L'Anthroponymie de Jean Molinet', 208–14; F. Rigolot, *Poétique et onomastique: L'Exemple de la Renaissance* (Geneva, 1977), 28–30.

[10] Brown, *Poets, Patrons, and Printers*, 157–66, and 'Éveil', 25–6, trace this evolution.

[11] Brown, 'Rise', 53–8, discusses the *Complainte de Grece* in this respect.

[12] The only other clear example of this technique in Molinet's work appears in his *Chroniques*, where Molinet mentions his *Naufrage de la Pucelle* while narrating events after the battle of Nancy (*Chroniques*, i. 209).

[13] These observations are indebted to Brown's analysis of the *Gaiges* in *Poets, Patrons, and Printers*, 220–4.

subject, Charles VIII's Italian expedition, and observes that he has composed the piece 'soubz termes bestiaux et poeticques' (277, ll. 11–12), thereby drawing attention to his depiction of Charles as a flying stag, hunted by the League of Venice.[14]

Molinet's correspondence also tends explicitly to present the *je* as a poet. Especially notable are the exchanges with his fellow *rhétoriqueur* Guillaume Cretin and the important patron Florimond Robertet (*FD*, ii. 834–41, 842–5); such pieces imply that Molinet forms part of a poetic community with a sense of its vocation.[15] As in Molinet's other verse, the self-consciousness of his correspondence is also manifested as formal virtuosity. Epistles often have the quality of puzzles, adopting structures which the addressee must decipher to make sense of the poem. Molinet addresses an octosyllabic *huitain* to Cretin in which, the recipient is told, the lines must be transformed into decasyllables by repeating the third and eighth syllables: thus 'Molinet sans bruit ne sans nom' becomes 'Molinet n'est sans bruit ne sans nom, non' (838, v. 1). Where the correspondence differs from much of Molinet's other work is in its explicit thematization of such devices: the *huitain* to Cretin, for instance, comes with the instructions 'Doublés le tiers et le huytiesme' (838). Furthermore, the epistles often have a competitive aspect, as poets attempt to outdo each other in ingenuity, and thereby to prove their worth: Cretin replies to Molinet with a *huitain* using the same structure (840).[16]

It is thus relatively uncommon, and principally in later poems, that self-consciousness in Molinet's poetry takes any form other than that of prosodic and linguistic complexity. As successive chapters will make clear, later *rhétoriqueurs* such as Lemaire and Bouchet foreground their activity in more diverse ways. Cynthia Brown has persuasively suggested that the later developments in Molinet's work are related to printing. His use of an increasingly

[14] See Devaux, *Jean Molinet*, 325–34. Devaux suggests that the prologue was designed to deflect potential criticism of Molinet's political stance in the *Voiage*, by implying the poem was written to order for a patron. On Charles's expedition, see Y. Labande-Mailfert, *Charles VIII et son milieu (1470–1498): La Jeunesse au pouvoir* (Paris, 1975), chs. 8–16.

[15] On Cretin and Robertet, see Cretin, *Œuvres poétiques*, ed. K. Chesney (Paris, 1932), pp. x–xxv; C. A. Mayer and D. Bentley-Cranch, 'Florimond Robertet: Italianisme et Renaissance française'. LeBlanc, *Va lettre va*, 99–104, discusses Molinet's verse epistles.

[16] See also LeBlanc, *Va lettre va*, 112. Thiry, 'Rhétoriqueurs de Bourgogne', 103–4, notes the political dimension in this exchange between the Burgundian Molinet and the French court poet Cretin.

prominent narratorial voice, which it is tempting to identify as the poet's own, developed as he questioned the subordinate position assigned him in the patronage relationship, and as he came to be involved in the printing of his work.[17] The more explicit signing of his poems was perhaps a response to the fate of his *Art de rhétorique*, printed in 1493 by Anthoine Vérard, who omitted Molinet's original dedicatory prologue containing his punning signature.[18] A related but more general tendency concerns the attribution of Molinet's poetry: earlier manuscripts contain fewer explicit attributions to Molinet than later volumes produced in a period when printing had begun to affect scribal practice.[19] It is important to set Brown's observations in a broader context, not only by considering the whole range of Molinet's self-conscious techniques across his poetic corpus, but also by examining the different ways in which scribes and printers may have understood and presented these techniques. A proper assessment of these issues must take account of texts' physical presentation, anthologization, and even textual variation across different manuscripts and printed volumes.

My examination of Molinet's poetry in manuscript form begins with the most important manuscripts, the two surviving anthologies of Molinet's verse. Where relevant, I compare individual poems' presentation in these volumes to that in the lost manuscript Tournai, Bibliothèque Communale, 105 (hereafter referred to as *A*, in accordance with the sigla allocated to manuscripts of Molinet's poetry by Noël Dupire).[20] I then examine significant single-text and multi-author manuscripts which foreground self-conscious aspects of Molinet's poetry, before discussing manuscripts which attenuate such aspects in favour of poems' historical

[17] Brown, *Poets, Patrons, and Printers*, 215–27, traces the narrative voice's evolution in Molinet's work.

[18] Ibid. 158–63. Two presentation copies contain a different prologue, attributing the *Art* to Henry de Croy: see M. B. Winn, *Anthoine Vérard, Parisian Publisher, 1485–1512: Prologues, Poems, and Presentations* (Geneva, 1997), 94–6.

[19] Brown, 'Du manuscrit', 105–7.

[20] This volume, the most important Molinet anthology, was destroyed in 1940. Detailed study is precluded, since the only available information about its physical characteristics consists of photographs published in *FD*, and descriptions in N. Dupire, *Étude critique des manuscrits et éditions des poésies de Jean Molinet* (Paris, 1932), 9–28, 113–14, and P. Faider and P. Van Sint Jan, *Catalogue général des manuscrits des bibliothèques de Belgique*, vi: *Catalogue des manuscrits conservés à Tournai (Bibliothèques de la Ville et Séminaire)* (Gembloux, 1950), 113–28.

or didactic value. In conclusion, I assess the prevalence of expressive scribal reactions to Molinet's sophisticated rhetoric and occasional thematization of writing, and reflect upon poems' attributions in manuscripts.

Manuscript compilations

The surviving anthologies of Molinet's poetry are Arras, BM, ms. 619, and BNF, ms. Rothschild 471, hereafter designated as *B* and *C* respectively, in accordance with Dupire's sigla.[21] The manuscripts are closely related: as Dupire noted,[22] they contain very largely the same pieces, which not only present close textual similarities but also appear in an almost identical order. A brief comparison of the manuscripts' constituent texts will underline the extent to which *B* and *C* resemble each other. In the interests of clarity and concision, I refer to the poems in each manuscript by the numbers which Dupire allocates them in his listing of the volumes' contents, *B* containing sixty-six such pieces and *C* eighty-four. Texts 1–18 in *B* are the same as texts 17–34 in *C*; similar correspondences exist between texts 20–5 in *B* and 35–40 in *C*, and 26–64 in *B* and 44–82 in *C*. Codicological analysis, moreover, reveals that most of the differences in these manuscripts' contents result from insertions made after the initial process of copying. *C* was apparently copied on a gradual, piecemeal basis, probably according to texts' availability. The bastard cursive in which most of the text is written displays some variation in letter forms, but not enough for distinct hands to be identified; rather, it suggests a single scribe's work over an extended period, during which some inconsistency is inevitable. Furthermore, the first three quires (fos. 1–26) are unsigned, whereas subsequent quires are signed in a sequence beginning at *a*; the unsigned quires were probably added to the volume after the sequence of the other quires had been fixed. Quires are of irregular size, often because they have been devoted to single texts, another indication of gradual copying. Two of Molinet's important *prosimetrum* allegories exemplify this tendency: the *Trosne d'Honneur*, a commemoration of Philip the Good composed shortly after his death in 1467, occupies an entire

[21] Dupire, *Étude*, 28–48, provides descriptions. [22] Ibid. 114–15.

quire (*FD*, i. 36–58; fos. 49–56), and the *Resource du petit peuple* of 1481, which reflects on the sorry predicament of the Burgundian Netherlands, takes up two quires (*FD*, i. 137–61; fos. 57–66). Of the material peculiar to *C*, texts 1–16 in Dupire's numbering appear on the first three quires, while texts 41–3 have probably been added to fill up the final page of a sequence of two quires otherwise devoted solely to the *Naufrage de la Pucelle* (*FD*, i. 77–99; fo. 88^v^). Similarly, texts 65–6 in *B*, of which the former is not represented in *C* and the latter corresponds to *C*'s text 3 (that is, a later addition), were apparently added after the rest of *B* was transcribed: they are copied in a hand different from that of the preceding material. Clearly, the manuscripts' contents in their original form are so similar that some kind of filiation between them cannot be doubted. However, some discrepancies in their contents cannot readily be explained by posterior insertion: text 19 in *B* does not appear in *C*, nor do texts 83–4 in *C* appear in *B* (though these, the final texts in *C*, were possibly transcribed at a later point). In the absence of more precise codicological information, the precise relationship between the volumes cannot be determined conclusively. One may have been copied from the other, either directly or via an intermediary manuscript; alternatively, both volumes may have been copied independently from a common source.

Issues of dating and localization underline the affinity between *B* and *C*. *C* bears three dates, apparently denoting the completion of different copying stints: '1526 3^{a} octobris', '1522 24^{a} nouẽbris', and '1520 13^{a} octobris' (fos. 16^{r}, 29^{r}, 66^{v}). The third date follows the *Resource du petit peuple*, and also appears after the *Resource* in *B* (fo. 43^{v}). Hence the dates in either *B* or *C*, or indeed both, may be inherited from an exemplar, whether this be the other of the two manuscripts or a third volume. The only certainty is that 13 October 1520 is the *terminus a quo* for at least part of both manuscripts: they thus indubitably post-date Molinet's death. *A* was probably also posthumous: it contained Molinet's last poem, the *Regrés de la mort Philippes, roy de Castille* (*FD*, i. 410–17: fos. 61^{r}–65^{r}).[23] The anthologies have links with Arras, a town of

[23] The *Regrés* was composed after 23 September 1506: Dupire, *Jean Molinet*, 37. *A* also contained poems by Philippe de Fenin, copied after Molinet's work (Faider and Van Sint Jan, *Catalogue*, 125–8); two of these were dated 1513 (fos. 434^{r}, 456^{r}), but it is uncertain whether this material was added by a later hand.

Austro-Burgundian allegiance in this period.[24] *C* belonged to the Arras chaplain Jean Garet, as is apparent from a note on fo. 201^{v}: 'Io. Garet Cappellanus me possidet Atrebateñ'. Garet's name also appears alongside the dates on fos. 16^{r} and 29^{r}, suggesting he was the original and not a later owner. *B* ends with an anonymous verse piece entitled *Le Mont Pelion* (fos. 171^{r}–193^{v}), added in a hand different from but broadly contemporary with that which transcribes most of Molinet's poems. This text praises the abbey of Mont-Saint-Éloi near Arras; a prose preface addresses 'Sire Mahieu du pre/religieux en la . . . maison du mont saint eloy' (fo. 171^{r}). The same name is mentioned in a note apparently indicating *B*'s owner and his device: 'Du present te garde | Matheus | De prato | Au futur regarde' (fo. 2^{r}). Significantly, *A* is also connected to Arras. *C* contains a loose sheet listing its contents, written in a hand different from that used elsewhere in the manuscript, whose heading begins 'Les nõbres en ciffres designent les feuilles de mõ grand Molinet escript a la main'. The leaf numbers added beside many of the texts listed coincide with these texts' positions in *A*; apparently Garet, or one of his successors, owned both *A* and *C*.[25] The major anthologies, then, are posthumous, and produced in an area where Molinet's Burgundian loyalties would find support.

B and *C* transmit both a large number and a wide variety of texts, encompassing the full range of functions and forms of Molinet's output. His encomiastic and occasional poems are well represented: all his important *prosimetrum* pieces on political subjects are included, though the *Complainte de Grece* is, as usual, reduced to its verse component.[26] But the manuscripts also contain many of Molinet's didactic, devotional, and bawdy poems, and prose and verse epistles. The poems present solely in *C* are mainly devotional, parodic, or correspondence pieces, and include several bilingual or formally intricate texts.[27] *B* thus contains relatively

[24] On Arras's loyalties, see J. Lemaire, 'Deux poèmes bourguignons sur la prise d'Arras en 1492: Un inédit de Molinet?'.

[25] *A*'s later owner was Jérôme Van Winge (*c.*1557–1637): see Faider and Van Sint Jan, *Catalogue*, 128; Académie Royale des Sciences, des Lettres et des Beaux-Arts de Belgique, *Biographie nationale* (Brussels, 1866–1985), xxvii. 350.

[26] The *Complainte* (*FD*, i. 9–26) is transmitted in full only in *A* (fos. 54^{v}–60^{v}) and Paris, Bibl. de l'Arsenal, ms. 3521 (fos. 186^{r}–193^{r}). On the latter manuscript, a 15th-century volume, see P. Michault, *Le Procès d'Honneur Féminin*, ed. B. Folkart, *MF* 2 (1978), 34–7.

[27] e.g. *Oroison a la Vierge Marie* (*FD*, ii. 483–90; fos. 17^{r}–19^{v}), a devotional poem whose stanzas each begin with a word of the Latin *Ave Maria*.

more political and occasional pieces, though the range of works it represents is no narrower than in *C*.

The order of material does not seem significant: there is no clear succession of genres or themes. More important is the range of texts presented: the scribes or owners clearly wished to provide the most complete anthology possible of Molinet's poetry. Many of the poems' titles explicitly mention Molinet's authorship. Consequently, it is perhaps unsurprising that no one area of Molinet's poetry is privileged by being concentrated at the beginning of the volumes; more relevant is the apparent desire for comprehensiveness. Breadth of output is foregrounded: the manuscripts provide a cross-section of the range of Molinet's talents. The resultant sense of a coherent *œuvre*, unified by the person of its writer, marks a departure from the practice observed in most fifteenth-century manuscripts.[28] It implies that the poet himself, not individual texts or themes, is the focus of interest. Scribes and owners treat Molinet as worth commemorating in posthumous anthologies; they thus acknowledge the claims to eternal renown for poets which several *rhétoriqueurs* articulate.[29]

Moreover, *B*'s and *C*'s scribes frequently highlight the linguistic and prosodic features of Molinet's verse. Neither volume is especially luxurious: both are paper manuscripts, transcribed *in campo aperto* without illustrations or decorations, excepting one rebus-poem in *C* which I discuss below.[30] Nevertheless, scribes use the page to establish a sophisticated grammar of legibility, which helps the reader distinguish between different textual functions or different languages, and which sometimes clarifies Molinet's prosody.[31]

C's layout is especially effective in this respect. Many poems' titles, and sometimes parts of *incipit*s or internal headings, are in a display script. Latin text in four pieces is in an approximation to humanistic script, with very long ascenders and overwhelmingly

[28] See N. F. Regalado, 'Gathering the Works: The *Œuvres de Villon* and the Intergeneric Passage of the Medieval French Lyric into Single-Author Collections', 94–7.

[29] See Jung, 'Simon Gréban', 78–84; C. M. Zsuppán, 'An Early Example of the Renaissance Themes of Immortality and Divine Inspiration: The Work of Jean Robertet'.

[30] Lemaire, *Introduction*, 106 n. 26, notes the lowly status accorded to manuscripts transcribed *in campo aperto*.

[31] M. B. Parkes, *Scribes, Scripts and Readers: Studies in the Communication, Presentation and Dissemination of Medieval Texts* (London, 1991), p. xv, defines the grammar of legibility as 'a complex of graphic conventions by which the written manifestation of language operates to facilitate access to the information it conveys'.

vertical letter forms.[32] In *Les Six triumphes en latin et franchois* (*FD*, ii. 584–7; fo. 16^{v}), frequent capital letters further enhance the script's Latinate quality (Plate 1.1). Latin and French text are also differentiated in other ways: in the Latin quatrains in *Complainte sur la mort madame d'Ostrisse* (fo. 43^{r-v}), for instance, even lines are indented. *A maistre David Walle* (*FD*, ii. 817–19; fo. 76^{r}) comprises alternate lines of Latin and French; all but one of the former are underlined. Furthermore, punctuation sometimes highlights prosodic devices, as when *virgulae* draw attention to the *rimes batelées* in *Le Trosne d'Honneur*:

> Ciel azure/Region aerine
> Aureine/splendeur reflamboyant
> Phebus phebe/et toutte estoille fine.
> (fo. 49^{v}; *FD*, i. 38, vv. 9–11)

B's grammar of legibility is less elaborate. There is no variation in script, excepting the use of a display script for the titles of the two poems added later, and the note identifying the owner. Consequently, Latin and French text are not always distinguished as effectively as in *C*. Of the poems where humanistic-style script is used for Latin text in *C*, the only piece where *B*'s scribe highlights Latin in any way is *Lettre a maistre Guerard de Watrellet* (*FD*, ii. 812–14; fos. 24^{v}–25^{r}), where Latin lines, which alternate with lines in French, are underlined. This apart, however, Latin and French are distinguished by the same methods as in *C*, and metrical punctuation used similarly. In both manuscripts, scribes also employ layout to highlight prosodic techniques used in specific poems. I shall examine three cases in detail, both to illuminate the poems' workings and to illustrate how Molinet's technical skill is presented.

In each manuscript, the rhyme-words in several poems are detached from the text and positioned to the right of the poem; this technique, the use of 'displayed rhymes',[33] has subtle implications for the reading of the poems. A striking example in *C* is 'Dame, j'ay sentu les façons' (*FD*, ii. 866–7; fo. 13^{r}), which initially appears to be a conventional love ballade but makes increasingly explicit use of military terms as sexual double entendres. The sexual sense is confirmed by the rhyme-syllables, 'cons', 'vis', and

[32] Fos. 16^{v}, 31^{r}–33^{r}, 47^{r}, 48^{v}.

[33] See Parkes, *Pause and Effect*, 100.

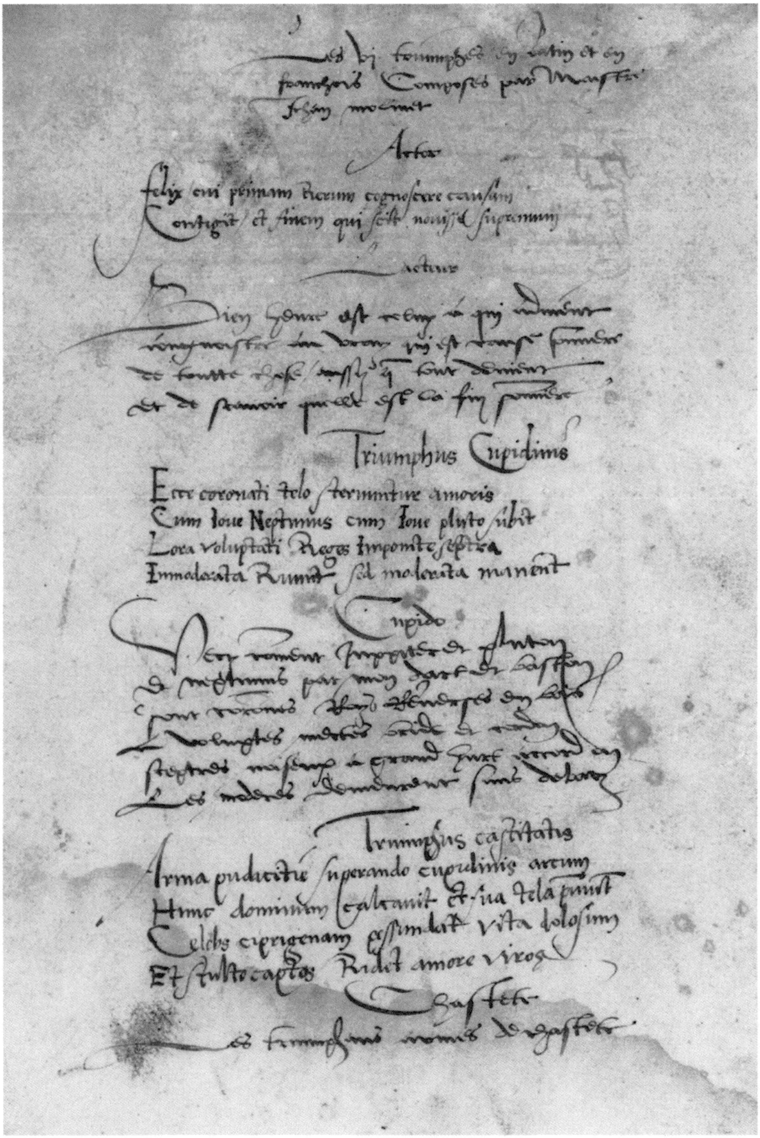

PLATE 1.1 Humanistic-style script for Latin text: BNF, ms. Rothschild 471, fo. 16v (Cliché Bibliothèque Nationale de France, Paris)

'cus'. In *C*, these syllables are detached to the right of the first stanza, and omitted altogether in subsequent stanzas (Plate 1.2). The displayed rhymes provide an early and obvious clue to the poem's double meaning, undermining the first stanza's apparently conventional courtly discourse and immediately signalling such innuendoes as:

> Se vous gaignés mon pont levis,
> Je clorray les deux tappecus. (vv. 21–2)

By fitting the rhyme-syllable to the verse with which it belongs, readers thus effectively invest the rest of the verse with a sexual sense. Indeed, the rhymes' absence after the first stanza requires readers to become complicit in the transgression of linguistic propriety, by supplying the required obscenities for each verse themselves. Readers are thus made acutely aware of their own interpretative activity. The rhymes' visual isolation reveals the rhyme scheme around which the poem has been elaborated, and thereby reminds readers that any poem is built up around a series of rhymes. Indeed, 'cons', 'vis', and 'cus' may be regarded as the basis not only of this text, in prosodic terms, but also of all love poetry, in thematic terms. Any amorous verse, by definition, is ultimately based upon sexual desire, upon the bodily parts which lyric conventions usually suppress via circumlocutions and euphemisms. This piece both reveals the true, unspoken preoccupation of love poetry, by detaching and therefore highlighting the rhyme-words, and hints that this preoccupation is normally denied direct expression. By reinforcing the tension between the poem's courtly opening and its developing bawdiness, the layout in a sense demystifies love poetry. In this respect, it indicates that the poem's self-consciousness lies not only in prosodic ingenuity, but also in its parodic aspect. By ironically commenting upon love poetry, and playing off it as the intertext for his poem, Molinet engages in metafictional parody.[34] The parodic love poem may itself be part of a well-established tradition,[35] but is no less self-conscious for it, especially when its comic reworking of amorous discourse is so suggestively presented on the page.

In *Ung present fait a l'empereur* (*B*, fo. 116^{r}; *C*, fo. 145^{v}), the name

[34] M. A. Rose, *Parody: Ancient, Modern, and Post-modern* (Cambridge, 1993), 91–9, discusses the relationship between parody and metafiction.

[35] See Johnson, *Poets as Players*, ch. 5.

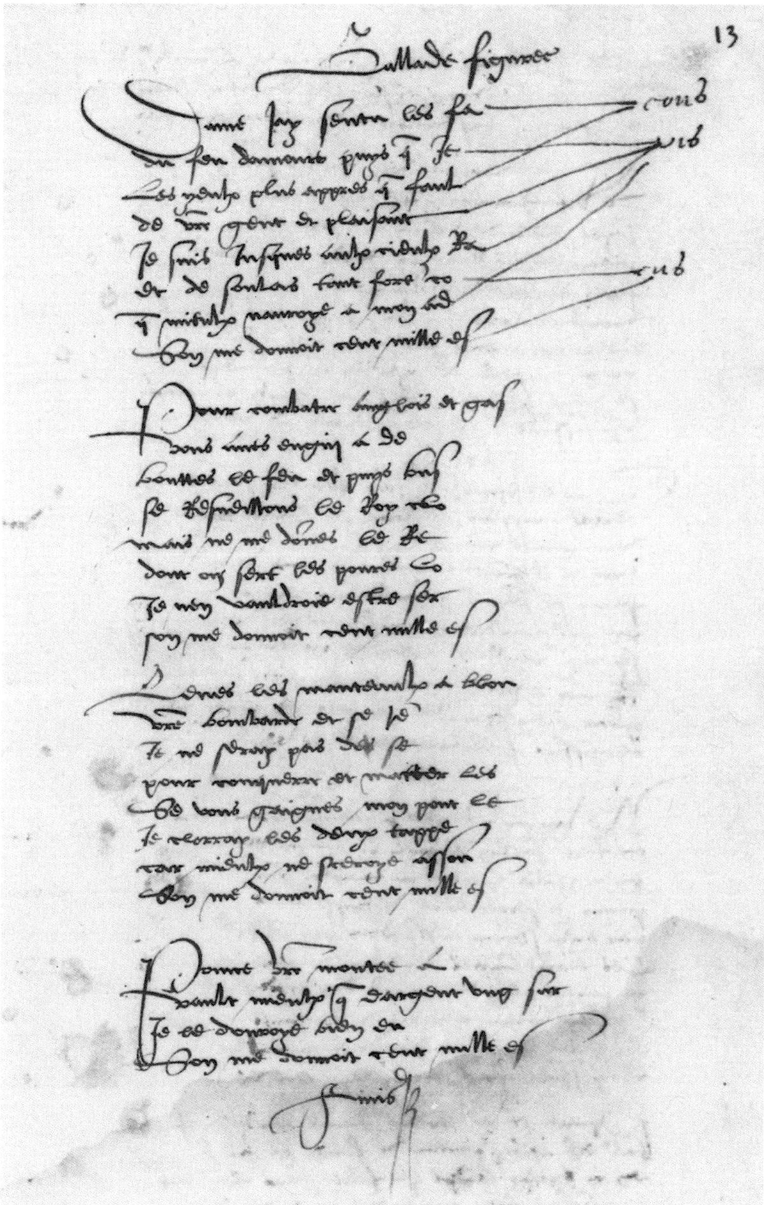

PLATE 1.2 Displayed bawdy rhymes: BNF, ms. Rothschild 471, fo. 13r (Cliché Bibliothèque Nationale de France, Paris)

of a bird forms each couplet's rhyme; these names are detached, and written once for each couplet. Each line also begins with a bird's name, usually highlighted by a *virgula* afterwards. The visual stress upon the names underlines their different, often simultaneous, values in the text. They are laudatory terms for Emperor Frederick III ('aigle', 'roy', 'duc', etc.); they are detached elements of other words, so that they both do and do not denote birds ('cyne' is simultaneously a swan and a part of the words 'macyne' and 'racyne'); they often have a metaphorical value (the eagle alludes to imperial power, the pelican suggests spiritual virtue); they also constitute a list of birds, independently of the poem's overall sense. Hence they may either be read 'horizontally', as part of the poem, or 'vertically', as forming two columns of birds' names, at the left-hand and right-hand edges of the text.

A's layout of this piece was even more suggestive, as the birds at both the beginning and end of each line were represented pictorially (fo. 159^{r}).[36] This requires a far more active role of readers, who must identify each bird and convert the pictures into names. The images sometimes permit easy recognition, via conventional attributes: the phoenix, for instance, is surrounded by flames. However, other images are not so clear: the phrase 'Divers oiseaux' (v. 16) cannot readily be inferred from the picture, which represents several birds. Consequently, readerly interpretations of the images are often provisional, and must be verified by other cues such as metre and sense. As a result, *A*'s readers must pay more attention to the birds' metaphorical significance than readers of *B* or *C*. One way in which readers may confirm their identification of a bird in an image is by assessing the metaphorical appropriateness of the bird they think may be represented. Certain birds' names, for instance, are likely to appear as epithets for Frederick III: 'aigle', 'roi', or 'duc' might be expected, whereas, say, 'geai' would not. The use of images, then, underlines the metaphorical potential of the birds' names, as might be expected in view of the medieval tradition of illustrated bestiaries which outline animals' allegorical significance.[37]

Moreover, *A*'s scribe was obliged to break the text into separate, widely spaced couplets, to provide enough space for the images.

[36] The image is reproduced as the frontispiece of *FD*, i.

[37] See D. Hassig, *Medieval Bestiaries: Text, Image, Ideology* (Cambridge, 1995).

Hence readers are made aware of the poem's materiality, of the requirements of book production. By thus presenting itself as overtly material, the poem reveals that it is indeed a 'present': it exists as a physical object, which can be sent and received as a letter or gift. Besides underlining the piece's formal and metaphorical complexity, then, *A* also draws attention to its reflexive self-commentary in the closing lines:

> te donne ce que j'ay,
> Divers oiseaux . . . (vv. 15–16)[38]

The only poem in *C* to use pictorial rhymes is 'Domine mi reverende' (*FD*, ii. 775–6; fo. 195^r; Plate 1.3).[39] The images, representing the rhyme-syllables 'dé' and 'table', and a shield bearing a death's head, are particularly rich in meanings for the whole poem. The poem refers to the 'escut' as a 'don', given in return for a gift 'Refraint d'ung ded et d'une table', in other words a poem using 'dé' and 'table' as rhymes (vv. 10–11, 4). Both the original gift and Molinet's response are depicted on the page, thereby stressing the text's physicality: both shield and poem are gifts. More importantly, the poem thematizes vision:

> Mon don qui est espouan [table]
> cest vng escu qui est bor [dé]
> de la mort horrible et doub [table]
> par dedens a miroir notable [*sic*]
> pour mirer visaige fon [dé]
> y bien penser est pourfi [table]
> quant vous laues bien regar [dé].
> (fo. 195^r; *FD*, ii. 775, vv. 10–16)

The shield is a mirror, and as such exemplifies a pervasive link between mirrors and mortality in medieval culture. Mirrors, it was thought, could show the dead future self of the viewer.[40] In *C*, then, the skull on the shield may be regarded as reflecting the reader's future self. The poem's *memento mori* theme is intensified, not only by the shield's mirror-like quality, but also more

[38] *A* also contained other pictorial rhymes, and numerous illustrations; see Faider and Van Sint Jan, *Catalogue*, 113–28. Consequently, its reader had a more challenging and rewarding visual experience than readers of *B* or *C*, and was inevitably made more sensitive to Molinet's language and prosody.

[39] In *B*, this poem lacks its rhyme-syllables (fo. 152^r). Presumably these were omitted to permit their eventual pictorial representation, as in *C*; however, no pictures have been provided.

[40] J. H. M. Taylor, 'Un miroer salutaire'.

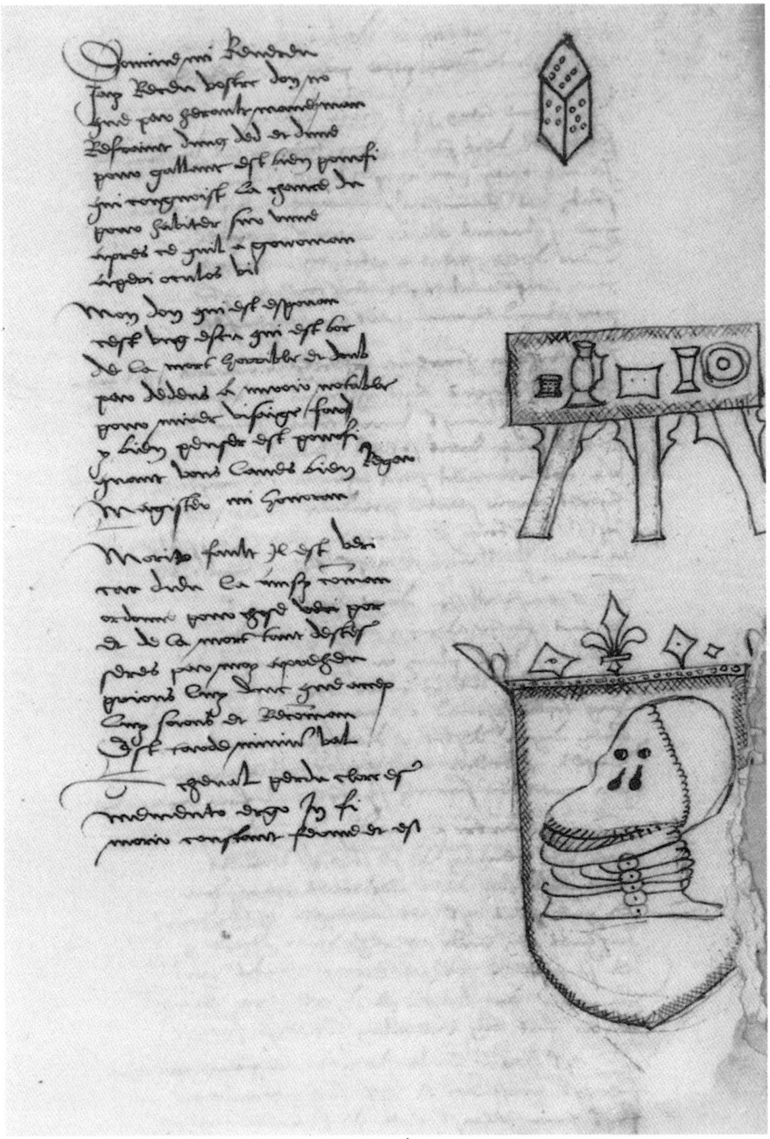

Plate 1.3 Pictorial rhymes: BNF, ms. Rothschild 471, fo. 195r (Cliché Bibliothèque Nationale de France, Paris)

generally, by the telling contrast between the skull and the die and table, attributes of transient worldly pleasure,[41] and by the emphasis which the pictures place on the poem's materiality. This is because textuality has a profound, traditional association with death, as it permits someone's words to be read in their absence, even posthumously.[42]

The construction of a complex grammar of legibility, and the tendency to highlight unusual formal features on the page, constitute two related facets of scribal practice, which displays notable prosodic and linguistic elements of Molinet's verse. Expressive layout not only has implications for individual poems' meanings, but also shows that scribes regarded Molinet's technical virtuosity as an important aspect of his work. The poems' sophistication requires scribes to develop appropriate strategies of presentation, to take visual charge of the texts and foreground their formal aspects for readers.[43] Displayed rhymes, for instance, require readers to play an active part in reconstructing a poem from the generative schema of rhymes on the page, and highlight both the poem's versification and the way in which Molinet often exploits rhyme-syllables' semantic value. This deciphering activity makes reading as self-conscious as writing: unusual layouts defamiliarize the reading process, and force readers to make choices as to how poems can be completed or 'solved'.

Significantly, scribal highlighting of formal features is not restricted to short pieces or correspondence poems. I have previously noted the distinction which *C*'s scribe makes between Latin and French in the *Complainte sur la mort madame d'Ostrisse*, an important occasional poem; similarly, displayed rhymes are adopted in several devotional poems,[44] while a stanza of *Le Naufrage de la Pucelle* is accompanied in *B* and *C* by a heading announcing that the verses may be read both forwards and backwards: 'Le premier met' est retrograde'.[45] Technical sophistication is acknowledged

[41] In *Lettre de recommanchon* (*FD*, ii. 808–9), Molinet employs 'dé' and 'table' as both rhyme-syllables and metonyms for a dissolute lifestyle (v. 30).

[42] See W. J. Ong, *Interfaces of the Word: Studies in the Evolution of Consciousness and Culture* (Ithaca, NY, 1977), 232–40.

[43] Layouts may be inherited from manuscript exemplars; nevertheless, they must originate in scribal practice at some point in the process of transmission. Whether *C*'s scribe or an unknown predecessor decided to present certain poems with displayed rhymes, the technique illustrates scribal sensitivity towards prosody.

[44] e.g. *Dictier des cinq festes Nostre Dame* (*FD*, ii. 450–2): *B*, fos. 117^{v}–118^{v}; *C*, fo. 148$^{r\ v}$.

[45] *B*, fo. 62^{v}; cf. *C*, fo. 83^{v}; *FD*, i. 90, vv. 1–8.

across Molinet's whole corpus, and often further underlined by titles which draw attention to prosodic structures. In *B*, for instance, the title of *Dictier des cinq festes Nostre Dame* reveals the acrostic which begins each stanza: 'Dictier a la benoiste vierge marie des cinq festes/Cõmenchãt chũn couplet O mater dei memẽto mei' (fo. 117ᵛ).

It is also important to note that these scribes' reactions to rhetoric, though far from unprecedented, go beyond what would normally be expected of vernacular poetry manuscripts in this period. While display scripts and different ink colours were frequently used for headings and Latin text, it is relatively rare for manuscripts to vary script within bilingual pieces, detach rhymes, or present them pictorially.[46] Hence the layout of Molinet's verse cannot wholly be explained in terms of contemporary practice; rather, the poems are treated on their own terms.

The Molinet anthologies, then, recognize the formal intricacy through which Molinet's self-consciousness is typically manifested. They also display an interest in Molinet as an individual, rather than in particular subjects or forms; it may be significant that they reproduce many of his correspondence pieces, which are both more self-referential and more 'personal' than many of his other works. While all three manuscripts apparently had connections with Arras, the strength of Molinet's literary influence in this area is not solely due to political allegiances. Pro-Burgundian propaganda is only one aspect of his output in these volumes; scribal interest in language and prosody is a more important consideration.

Self-consciousness in other manuscripts

It is perhaps inevitable that the anthologies of Molinet's verse should stress its language and versification, since their compilers

[46] For general tendencies in manuscript layout, see G. Hasenohr, 'Traductions et littérature en langue vulgaire', 289–315. Various anthologies containing Molinet poems illustrate that Latin and French are not often visually differentiated. The 16th-century volume BNF, ms. fr. 24038 (see Dupire, *Étude*, 70) uses display script for headings, but makes no linguistic distinction on the page. The same is true of Mons, Archives de l'État, ms. divers 114 (transcribed between 1529 and around 1600), which nevertheless includes several short poems with displayed rhymes: see A. Louant, *Le 'Livre de ballades' de Jehan et Charles Bocquet, bourgeois de Mons au XVIᵉ siècle* (Brussels, 1954).

are so obviously interested in Molinet and the aesthetic he represents. Other manuscripts containing his works are less likely to reflect such priorities: volumes comprising a single Molinet piece may have been produced out of interest in the text's subject-matter, while multi-author collections which include Molinet poems may have been compiled on thematic lines, with little interest in authors' identity. In the remainder of this chapter, I shall examine the place of these two types of manuscript in the contemporary reception of Molinet's self-conscious techniques. The current section analyses three manuscripts which underline such techniques; I study two volumes in which diverse scribal strategies are employed to present a wide range of Molinet texts, before focusing upon the only extant illustrated manuscript of any of Molinet's *prosimetrum* pieces. In the final section, I examine manuscripts which privilege poems' historical or political themes at the expense of their self-conscious aspects.

An important manuscript, for both the quantity of Molinet poems it transmits and its use of layout, is Cambridge, Gonville and Caius College, MS 187:220. Some seventy Molinet poems occupy over two-thirds of the manuscript, a small paper volume transcribed largely in the early 1520s by one Jean Hanon and his son Arturus.[47] The former, especially, establishes a sophisticated grammar of legibility. Headings are usually in a display script in red ink, as are some *incipits*; stanzas or sections are marked by red *lettres de tourneure*; *virgulae* or *punctūs* frequently mark verse caesuras, consequently highlighting internal rhymes where they occur. Such techniques draw attention to poems' formal organization and versification. Visual distinctions between Latin and French are less common, but are occasionally drawn.[48] Several poems have displayed rhymes or other expressive layouts; this practice characterizes occasional pieces as well as short or ludic poems. In the *Resource du petit peuple*, for instance, Jean Hanon draws attention to a sequence of alexandrine couplets with which Molinet adorns a

[47] Jean transcribed pp. 1–464, Arturus the remainder; there are some brief notes in other hands. See K. M. Munn, *A Contribution to the Study of Jean Lemaire de Belges: A Critical Study of Bio-bibliographical Data, Including a Transcript of Various Unpublished Works* (Scottdale, Pa., 1936), 183–8. The manuscript's contents have not previously been adequately catalogued: I shall examine the volume fully in a future study.

[48] e.g. in *Les Regrés de la mort Philippes* (pp. 408–14), where most of the Latin verse is in two columns of display script.

passage of prose, by writing most of the alexandrines on separate lines (p. 312; *FD*, i. 139, ll. 69–74).[49]

Layout is used rather differently to emphasize the punning rhymes in Molinet's lament on syphilis, *Complainte d'un gentilhomme a sa dame* (*FD*, ii. 729–31; pp. 454–7). In this piece, the first line of each couplet ends in a compound word beginning with the syllable 'con', whereas in the second line 'con' has an explicit sexual meaning:

> Ah! la belle, pour qui plus de maulx je comporte
> Que pour femme aujourd'huy qui sus terre con porte . . . (vv. 1–2)[50]

Jean both detaches the rhyme-syllables and highlights them in red ink; moreover, he represents 'con' by two distinct graphic forms, which correspond to its two values in the poem. As a bound morpheme, in the odd lines, it is written 'cõ'; when it has an explicit sexual meaning, in the even lines, it is abbreviated to '9'. This distinction visually accentuates the way in which the poem alternately reveals and conceals obscenities.[51]

A smaller but similarly varied selection of Molinet's verse appears in BNF, ms. fr. 19165 (referred to by Dupire as *h*), which consists of three quires of Molinet's poetry bound before a fifteenth-century copy of Adenet Le Roi's romance *Cléomadès*.[52] Besides including various bilingual texts and puzzle- and rebus-poems,[53] *h* also reveals scribal interest in the language and metaphors of Molinet's occasional verse. In the second stanza of the epitaph *Au comte Englebert de Nassau* (*FD*, i. 394–8), Molinet lists the virtues which the Burgundian diplomat embodied, virtues beginning with the successive letters of Engelbert's name (Espoir, Nobilité, etc.).[54] The scribe picks out the virtues' initial letters in large gilded *lettres de tourneure* (fo. 14ᵛ); he thereby both visually glorifies Engelbert and highlights Molinet's ingenuity.

[49] See A. Armstrong, 'The Practice of Textual Transmission: Jean Molinet's *Ressource du petit peuple*', 272, 274. This rhyming prose is also highlighted in BR, ms. 21551–21569, a composite manuscript copied by three hands. The third scribe has transcribed occasional poems by Molinet alongside several pieces commemorating the French defeat at Pavia in 1525 (ibid. 279; Dupire, *Étude*, 76–7); the alexandrines in the *Resource* are underlined (fo. 170ᵛ).

[50] On this poem, see Cornilliat, *'Or ne mens'*, 67–8, 72–4, 271–7.

[51] See Johnson, *Poets as Players*, 277–8.

[52] Dupire, *Étude*, 67–70.

[53] For examples, see A. Armstrong, 'Two More Rebus-Poems by Jean Molinet?'

[54] On Engelbert, see Devaux, *Jean Molinet*, 144–6, 539–43.

In the *Voyage d'Espaigne* (*FD*, i. 373–80; fos. 1^r–5^r), which recounts a tour of European courts by Philip the Handsome, the protagonists' coats of arms are portrayed above or below the stanzas which first mention them, on backgrounds in the form of heavenly bodies which thus evoke the poem's governing astrological metaphor.[55] The arms are those of Philip and his wife Joanna, represented in the poem by the sun and moon respectively; Engelbert of Nassau (represented by Saturn); the French monarchy (Venus); Ferdinand of Spain (Mars); Emperor Maximilian (Jupiter); and Jérôme Lauwereyn, lord of Watervliet, Philip's treasurer-general (Mercury). These shields, and the poem's title, suggest to readers the true subject of the astrological narrative, provoking them into trying to identify the figures. Yet the shields also show the distance between the historical figures and the planets representing them: coats of arms are manifestly out of place in the realm of astrology, as are planets in that of heraldry. The shields are a kind of middle term, part of neither the narrative's literal, astrological level nor its allegorical, political level (which they merely signify, but do not constitute). By accentuating the gap between these levels, they encourage readers to reflect upon the reasons why each planet has been used to represent each protagonist.

These reflections reveal certain overtones in the metaphors. Using the sun and moon to stand for Philip and Joanna exalts the former at the expense of the latter, since the moon receives its light from the sun: hence the specifically Burgundian side of the alliance is privileged above the Spanish side. As for Venus, her status as goddess of love renders her a fitting embodiment of the French monarchy in a poem celebrating peace and reconciliation between France and Burgundy. The illustrator, however, has added a canton argent to the French arms (fo. 1^v). Normally a *brisure*, this charge may have a tacitly pejorative significance here, suggesting that France's arms are not quite as whole, and therefore less worthy, than Burgundy's.[56] While Molinet's text promotes international understanding, then, its physical actualization in *h* evinces a more narrowly pro-Burgundian stance on the illustrator's part.

[55] On this poem, see ibid. 339–40, 515, 585–6.

[56] I am grateful to Jean Devaux and Michel Pastoureau for their suggestions as to the canton's meaning.

The Cambridge volume, and manuscript *h*, show the extent to which scribes acknowledge the formal self-consciousness of Molinet's poetry; however, they provide no instances of illustrators' reactions to the texts. Such reactions are abundant in BNF, ms. fr. 14980 (referred to by Dupire as *E*), to which I now turn my attention. The volume is entirely occupied by *Le Naufrage de la Pucelle* (*FD*, i. 77–99), a *prosimetrum* allegory from mid-1477 recounting the tribulations of Mary of Burgundy, and the Burgundian Netherlands, after the battle of Nancy.[57] The death at Nancy of Charles the Bold, Mary's father, left the House of Burgundy without a male heir; it was followed by a French invasion, and widespread civic unrest.[58] In the *Naufrage*, these events are presented as the vision of a first-person narrator, or *acteur*, as such narrators were conventionally termed in this period. Mary is represented as the Pucelle, left in charge of a ship denoting Burgundy. She has three allegorical companions: Coeur Leal, who embodies Burgundy's loyal retainers; Noblesse Debilitee, personifying unreliable aristocrats who wavered between Burgundy and France; and Communauté Feminine, representing the third estate, prone to violence yet unswervingly faithful to the Pucelle.[59] The Pucelle initially appeals for help to God, the Virgin Mary, and Emperor Frederick. Coeur Leal consoles her, and navigates the ship, helped or hindered by his companions. A whale, representing Louis XI, first attempts to swallow the ship, then sends a dolphin to charm the crew, accompanied by two sirens which denote French propaganda. This episode alludes to Louis's hope that Mary would marry the French Dauphin, so that the Burgundian Netherlands would revert to France. Eventually, the whale resorts to force, sending an army of sea-monsters to attack the ship. The crew are rescued by an eagle, representing Frederick, who sends the Pucelle its son (denoting the future Emperor Maximilian, whom Mary was to marry). Molinet's text

[57] On the *Naufrage*'s date, see Devaux, *Jean Molinet*, 266–9.

[58] On this period, see J. Devaux, 'Les Soulèvements urbains de 1477 sous le regard des chroniqueurs du temps'; W. P. Blockmans (ed.), *Le Privilège général et les privilèges régionaux de Marie de Bourgogne pour les Pays-Bas, 1477* (Courtrai, 1985). On Molinet's treatment of events after Nancy, see Molinet, *Chroniques*, i. 172–213; J. Devaux, 'Le Rôle politique de Marie de Bourgogne au lendemain de Nancy: Vérité ou légende?', 'Les Guerres de Louis XI au *pays de Haynau*: Le Témoignage de Jean Molinet'.

[59] On these figures, see Devaux, 'Rôle', 391–4; *Jean Molinet*, 361–4, 394–5, 399–406, 431, 463–4.

may be regarded as an attempt to solicit support for Mary's Habsburg alliance, and to exhort Burgundian nobles to resist Louis's aggression.[60]

E is the only extant manuscript to contain the *Naufrage* alone; a small volume, probably dating from the fifteenth century,[61] it is well produced on parchment, with headings in blue ink, foliate *lettrines*, and ten pen-and-ink illustrations by an unknown artist. These illustrations influence readerly perceptions of three significant features of the *Naufrage*, which I shall examine in turn: its use of political allegory; the presence of elements linking the Pucelle with the Virgin Mary, as well as Mary of Burgundy; and Molinet's manipulation of the *acteur*, which gives him, as author, an elusive presence in the text.

The images appear immediately before the passage they depict, and generally follow textual details closely. The description of the way in which the Pucelle's companions navigate the ship, for example, is faithfully rendered in an illustration on fo. 10v (Plate 1.4): 'Coeur leal . . . regentoit l'office des pillos, donnoit regard au pol artique et, speculant en la quarte marine, commandoit a sonder. Noblesse debilitee demonstroit semblant d'estre bien empeschee, mais gueres ne se traveilloit, sinon de planer oultre et de caller le voille. Communauté suoit de paine autour des ancres . . .' (87–8, ll. 130–7). The close similarity between text and illustration here provides a useful starting-point from which to consider Molinet's allegory. What is depicted is the literal, not the allegorical, action: we see a ship, a whale, a dolphin, rather than Burgundy, Louis, or the infant Charles VIII. These literal depictions reflect the way in which Molinet's narrative functions literally as well as allegorically. While the various characters and motifs have a political meaning, their literal meaning has its own value, often providing further connotations which nuance the political allegory; the sirens, for instance, aptly suggest the threatening untrustworthiness of French blandishments. By making the literal sense visible, *E*'s illustrations underline that Molinet's allegory cannot be read on a purely historical level, as a veiled representation of current affairs.

[60] Devaux, *Jean Molinet*, 266–9; 'Rôle', 401.

[61] Fo. 8v bears a catchword written vertically, characteristic of 15th-century manuscripts: see Lemaire, *Introduction*, 176.

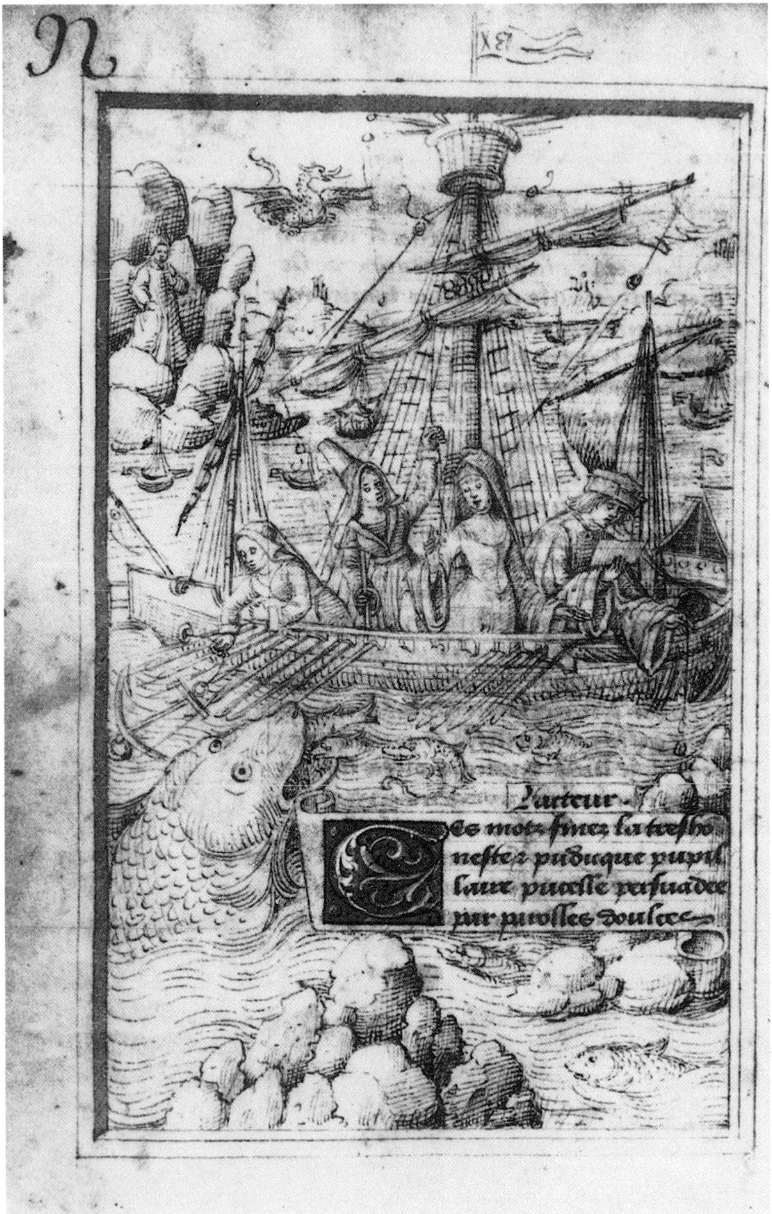

PLATE 1.4 A whale attacks the Pucelle, assisted by a dolphin: BNF, ms. fr. 14980, fo. 10[v] (Cliché Bibliothèque Nationale de France, Paris)

In one respect, *E*'s illustrations depart from textual detail, and depict an allegorical rather than literal motif; yet this departure paradoxically confirms the importance of Molinet's literal narrative. The final two images depict a knight, with an eagle on his shield, in the Pucelle's ship (fos. 18^{v}, 19^{r}; Plates 1.5, 1.6). The text which these drawings illustrate recounts that, to help the Pucelle, the eagle sent her his son, a 'jonne aigle ardu et vaillant' (98, ll. 71–2).[62] However, while Molinet represents Frederick and Maximilian as eagles, the illustrator has partly solved Molinet's allegory; Maximilian is presented in human form, bearing the imperial eagle on his shield. This visual interpretation of the narrative makes explicit the young eagle's metaphorical relationship to Maximilian. Readers must equate the knight in the illustration with the bird in the text, neutralizing the inconsistency between text and image by establishing that knight and bird have a common referent. Yet it is precisely this discrepancy which underlines the interpretative leap readers must make, from literal to allegorical reading. In this instance, we must take account of both the literal and the allegorical sense of the young eagle; the former cannot be ignored.

Molinet thus employs coherent, suggestive metaphors, not ciphers which may be discarded once their political referents have been identified. The allegory's connotative richness is especially clear in the *Naufrage*'s depiction of the Pucelle, who has traits reminiscent of the Virgin as well as Mary of Burgundy. Readers concentrating upon the text's political allusions may miss the Marian symbolism; however, *E*'s illustrations call attention to it, further underlining the sophistication of Molinet's allegory.

The complex of connotations linking the Pucelle with the Virgin is established from the outset by the very name 'Pucelle', which defines the heroine through her virginity. Once readers realize that the Pucelle represents Mary of Burgundy, a further connection is apparent, in that the princess and the Virgin have the same name. This name, moreover, is phonically linked to the Pucelle's virginity when the latter laments her plight in verses characterized by striking *annominatio*:

[62] This passage does not appear in *E*: five leaves have been lost from between fos. 9–10, 12–13, 13–14, and 18–19.

PLATE 1.5 The climactic battle in *Le Naufrage de la Pucelle*: BNF, ms. fr. 14980, fo. 18v (Cliché Bibliothèque Nationale de France, Paris)

PLATE 1.6 Author and narrator within the same picture: BNF, ms. fr. 14980, fo. 19r (Cliché Bibliothèque Nationale de France, Paris)

A marier, sans mari suis Marie,
Triste, marrie, en point de mariage . . .
A pou diray ung Ave Maria;
La dame est bien qui ung bon mari a. (81, vv. 17–18, 23–4)

Indeed, in the following stanza, the Pucelle prays to the Virgin herself, confirming the affinity between them:

Vierge Marie et estoille marine,
Mere et marrine, a toy me rendz encline. (81, vv. 27–8)

The link is intensified when Coeur Leal consoles the Pucelle after her lament, addressing her in strongly Marian terms: 'O tres humble et tendre virgine, lys odorant' (84, l. 9).[63]

E's illustrations reflect this imagery in varying ways. The Pucelle's lament is accompanied by a drawing representing the Virgin in one of her traditional guises, as *stella maris*, the 'estoille marine' of the Pucelle's prayer (fo. 4^{v}; Plate 1.7).[64] Consequently, the Virgin's prominence is clearly underlined, rendering it more likely that readers will perceive the link between her and the Pucelle. A less direct visual reflection of the Marian imagery appears in Plate 1.5, which depicts the eagle in a diving position next to the ship. The combination of bird and ship recalls the dove's return to the Ark, a connection reinforced by the text's reference to the Pucelle's ship as a 'precieuse arche' (77, l. 16). But the eagle's diving movement, and the evocation of Noah's dove, also recall, in conjunction, the dove of the Holy Spirit, typically represented in Annunciation images as diving towards the Virgin.[65] While it may seem paradoxical for an eagle to evoke a dove, the equivalence is suggested by the density of religious connotation elsewhere in the text. Furthermore, the link between eagle and dove has appropriate resonances for Frederick's historical role, as bringer of peace to the Burgundian Netherlands. In both these images, the illustrator's use of conventional visual models (the diving bird recalling the dove, Mary as *stella maris*)

[63] Molinet later explored Mary's affinities with the Virgin in *Le Chappellet des dames*, which claims that 'nostre vertueuse princesse Marie est la vive face, protraction et imaige de la seulle emperis du ciel' (*FD*, i. 125, ll. 156–7). His account of Mary's marriage to Maximilian in the *Chroniques* also adopts Marian imagery, as well as the eagle metaphor used in the *Naufrage*; *Chroniques*, i. 224–37. See Cornilliat, *'Or ne mens'*, 610–18; M. Randall, *Building Resemblance: Analogical Imagery in the Early French Renaissance* (Baltimore, 1996), 46–53.

[64] See E. Kirschbaum (ed.), *Lexikon der christlichen Ikonographie* (Rome, 1968–76), iv. 214–15.

[65] L. Réau, *Iconographie de l'art chrétien* (Paris, 1955–9), II. ii. 185.

PLATE 1.7 The Pucelle prays to the Virgin: BNF, ms. fr. 14980, fo. 4v (Cliché Bibliothèque Nationale de France, Paris)

suggests allegorical equivalences which highlight the Marian imagery.

Hence the illustrations again underline that the text is not simply a disguised account of political events. Molinet is not writing history, but absorbing historical events into literature, and transforming them into elements which signify within an intricate textual structure. The drawings indicate a way of reading the *Naufrage* which avoids interpreting its metaphors univocally: the visual juxtaposition of the Pucelle with the Virgin reminds us that the Pucelle, and through her Mary of Burgundy, is endowed with Marian qualities, while the depiction of Frederick as a diving eagle indirectly accentuates the allegory's religious dimension. Molinet's poetic ingenuity is thereby underlined: he has not merely provided masks for a historical narrative, but created sophisticated metaphorical constructions. However, his writerly activity is foregrounded even more clearly by the way in which the illustrations highlight the *acteur*'s role in the *Naufrage*, and his relationship to Molinet as the text's historical author.

Molinet's narrator frames the action of the *Naufrage*, introducing it as a vision, and closing it with a stanza outlining the writing and dedication of the text:

> Ainsy rimant sur la marine,
> Escripvis le doulx et l'amer,
> Sans flater parin ne marrine,
> De ce que je vis en la mer;
> A celle que je voeul amer
> Le presentay en sa naicelle:
> C'est, qui ainsy le voeult nommer,
> Le Naufraige de la pucelle. (99, vv. 17–24)

Between the introduction and conclusion, the narrator relates events in the third person, so that his role in mediating the action to the reader is less prominent.

E's illustrations render the *acteur* rather more prominent than the text alone; he is present in every image. We are thereby reminded that the narrator inevitably interposes himself between the events and the account of those events presented to the reader. Moreover, the presence of the narrator evokes that of Molinet: the use of first-person narration inevitably conflates the narratorial *je* with the author who writes that *je*.[66] By keeping the narrator before our eyes, the images intensify this effect: they constitute visual reminders of a figure we are tempted to identify, if only partially and provisionally, with Molinet himself. As the *Naufrage* in *E* is not attributed to Molinet (at least not in *E*'s present state), readers may not know the author's identity; nevertheless, the illustrations help draw attention to his relationship with the *acteur*. More insistently than the text alone, they underline that it is the poet's activity which has created the *Naufrage*.

Indeed, in the final illustration (Plate 1.6), the focus shifts definitively from the allegorical narrative onto the writing of this narrative. In the foreground, the narrator writes down what he sees; in the background, the author presents his poem to his patron, on board the Pucelle's ship. The image responds to a hint in the final stanza, where the *acteur* states that he presented his work to a lady 'en sa naicelle'. This clearly lends itself to the illustrator's interpretation, according to which the author dedicates his poem, not to his real patron in the world outside the text, but to one of the allegorical protagonists. Two distinct narrative frames coexist within a single pictorial frame: in metaleptic fashion, both author and narrator are represented simultaneously within the nautical scenario of the allegorical action.[67] Indeed, the narrator witnesses and writes down the author's activity, accentuating the way in which Molinet's reference to his work's presentation is pictorially interpreted as forming part of the narrative.[68]

As a result, this image highlights the metaleptic quality of Molinet's final stanza, where the referent of *je* appears to oscillate between the first-person narrator and the poet himself. We are obliged to consider the extent to which this stanza refers to Molinet's own writerly activity, and to which the *acteur* in the *Naufrage* may be identified with Molinet. Easy distinctions between

[66] Émile Benveniste differentiates linguistically between these two subjects in 'La Nature des pronoms'.

[67] Gerald Prince defines metalepsis as 'the mingling of two distinct diegetic levels': *A Dictionary of Narratology* (Lincoln, Nebr., 1987), 50.

[68] The doubling of narrator-figures could be interpreted as a device showing the passage of time, whereby the narrator writes his poem in the foreground before presenting it in the background. However, the fact remains that the poem is presented within its own fictional world, so that the image incorporates extratextual, authorial activity into the narrative.

author and narrator are problematized, in a visual play of identity and difference between the two. The illustrator further complicates the issue by depicting the text's dedicatee, whose identity Molinet does not explicitly specify, as the Pucelle. To regard her as standing straightforwardly for Mary of Burgundy, and hence to interpret the presentation scene as a coded representation of the text's historical dedication, would be simplistic, for the illustrations have shown throughout the inadequacy of such univocal, referential readings. Indeed, the image shows the presentation as taking place within the space of the allegory; in presenting his work, the poet steps not outside but inside his fiction, into a 'naicelle'. Diegetic levels are blurred, so that it becomes impossible to make rigid distinctions between the allegorical protagonists, the *acteur*, and the historical figures Molinet and Mary. Molinet's presence in the *Naufrage* is rendered equivocal and elusive.

E's illustrations thus draw attention to the *Naufrage*'s self-conscious aspects, by according them greater prominence than does the text alone. They accentuate both the sophistication of Molinet's allegory, with its narrative coherence and religious connotations, and the ambivalent relationship between Molinet and the *acteur* which hints suggestively at the poet's compositional activity. Ultimately, the images make it clear that the *Naufrage* is no mere *texte-à-clef*; rather, it presents a complex field of interrelated metaphors and elusive first-person voices.

Attenuations of self-consciousness

While the manuscripts discussed above foreground Molinet's formal virtuosity, use of allegory, and allusions to the writing process, their scribes' and illustrators' practice does not necessarily render readers oblivious to the political and didactic orientation of much of the poetry. Other volumes, however, lay rather more explicit stress upon this orientation, so that poems' self-conscious features assume lesser importance. Two such cases deserve especial examination, as particularly interesting illustrations of significant tendencies in the transmission of Molinet's poetry.

Six pieces by Molinet, primarily occasional in purport, are

transmitted in BNF, ms. fr. 2200 (referred to by Dupire as *e*).[69] Other material in *e* is largely pro-Burgundian, and often virulently anti-French; indeed, several poems celebrate the victory of Charles V's imperial army over French forces at Pavia in 1525, seen as a Burgundian triumph.[70] The title of one such piece amply illustrates the way in which Pavia was regarded as revenge for past French misdeeds, in this case Charles VIII's repudiation of Margaret of Austria in favour of Anne of Brittany: 'La complainte du roy francoys qui se chante sur le chain de la chanson faicte de ma dame marguerite de flandres quãt elle fust delaissie pour prandre bretaĩgne' (fo. 44r). Hence, well before readers come to Molinet's poems (which are among the last pieces in *e*), a clear context, or 'codicological horizon',[71] is established: this horizon encourages a reading which privileges Molinet's political allegiances and the texts' referential content, rather than their sophisticated rhetoric.

e is unusual among manuscripts of Molinet's work, in that it was apparently compiled in Burgundy proper: several pieces concern aspects of Dijon's history.[72] In this light, it is significant that Molinet's pieces are all attributed, whereas other material in *e* is anonymous. Molinet evidently enjoyed a reputation as a Burgundian patriot which extended beyond the Netherlands and north-eastern France. *e*'s regional affiliations are underlined by the heading provided for *Le Testament de la Guerre* (*FD*, ii. 718–24), in which Molinet lists war's effects in the form of bequests before abruptly subverting the testamentary fiction to reveal that war is still very much alive: 'Le testament de la guerre compose ꝑ maistre Iehan molinet apres q̃ la reduct' de salins ⁊ de la conte de bourg.ne fut faicte' (fo. 81r). It is uncertain which conquest is being referred to, and whether this indeed provoked the *Testament*'s composition;[73]

[69] Dupire, *Étude*, 55–6; an unpublished *notice* of this manuscript is held in the IRHT.

[70] For an account of the battle, see R. J. Knecht, *Renaissance Warrior and Patron: The Reign of Francis I* (Cambridge, 1994), ch. 11; for literary reactions to Pavia, see C. Thiry, 'L'Honneur et l'Empire: A propos des poèmes de langue française sur la bataille de Pavie'.

[71] The term is coined in L. de Looze, *Pseudo-autobiography in the Fourteenth Century: Juan Ruiz, Guillaume de Machaut, Jean Froissart, and Geoffrey Chaucer* (Gainesville, Fla., 1997), 79, to refer to processes of composition. However, it aptly expresses the interpretative guidance provided by the material context in which texts are anthologized.

[72] One heading, for example, reads: 'Nota bene de la maison des ducz de bourg.ne estant en leur ville de dijon a pñt occupee' (fo. 6v). Other references to Dijon appear on fos. 103r and 104v, among others.

[73] See Devaux, *Jean Molinet*, 480–6.

however, the heading clearly alerts readers to the poem's relationship to local historical events, rather than Molinet's ingenious manipulation of the testamentary form.

It is sometimes possible to adduce quite specific links between the content of Molinet's material and the preoccupations of *e*'s compiler. An interesting case is that of *La Tres desiree et prouffitable naissance de Charles d'Austrice* (fos. 95r–100r; *FD*, i. 352–8), dating from shortly after Charles V's birth in 1500. Here, Molinet not only celebrates Charles's birth, representing the infant and his lineage as the 'arche de paix' (v. 1), but also draws attention to his compositional activity through intricate versification and references to his work.[74] In *e*, the poem takes on a prophetic quality, as events at Pavia prove Molinet's predictions of Charles's future glory to be accurate:

> Charles sera vr̃e vray heritier
> en ce quartier cler dhõneur pur et mũde
> charles sera puissant a souhaitier
> pour nous aydier contre mortel dangier
> fort cõme ogier/ lung des plus fort [*sic*] du monde
> sur mer ꝑfunde/ sur fleuue τ sur vnde
> sur terre ronde aura triũphant nom.
> (fo. 99r; *FD*, i. 357, vv. 145–51)

It is this prophetic value, rather than the piece's various self-conscious features, which is most prominent.[75]

In several manuscripts, the historical subject-matter of Molinet poems is even more prominent, as the pieces are transcribed explicitly for their documentary value. This tendency is strikingly exemplified in BNF, ms. fr. 24052 (referred to by Dupire as *j*), which dates from 1507–8, and largely comprises historical material relating to Tournai, a town of French allegiance during Molinet's lifetime.[76] Three of Molinet's occasional pieces appear among a sequence of documentary texts following Jean de

[74] See Devaux, *Jean Molinet*, 512–15; Brown, *Poets, Patrons, and Printers*, 224–7. Molinet also uses the metaphor of an ark or house with pillars, denoting a dynastic house supported by its scions, in his *Chroniques*, in the prefaces (i. 25–8; ii. 591–3) and upon Charles the Bold's death (i. 207).

[75] For further analysis of Molinet material in *e*, see Armstrong, 'Practice', 278–9. The historical and political orientation of Molinet's poetry is similarly privileged in two other manuscripts containing Burgundian reactions to Pavia, BR, ms. 21551–21569, and Mons, Archives de l'État, ms. divers 114 (see above, nn. 46, 49).

[76] Molinet violently attacked the town in his *Dictier sur Tournay* (*FD*, i. 181–92); see Devaux, *Jean Molinet*, 317–19.

Nicolay's *Kalendrier de la guerre de Tournay*, which occupies most of *j*.[77] Introductory rubrics not only situate these poems in their historical context, but also criticize Molinet's pro-Burgundian attitudes. The introduction to *Le Bergier sans solas* (fos. 368ᵛ–376ʳ; *FD*, i. 209–24), for instance, links the poem to the Peace of Arras of 1482, and implies that Molinet's professed concern for peace is hypocritical in the light of Archduke Maximilian's disrespect for that treaty: 'Sensieult aultre oeuure dud' mollinet intulle [*sic*] les bregiers sans soullas lesquelx ilz [*sic*] fist trois [ans] apres la paix peu auant que le duc doctrisse [*sic*] eult anfraint ycelle' (fo. 368ᵛ).[78]

Rather more direct criticism is directed at Molinet's *Journee de Therouenne* (fos. 344ʳ–348ᵛ; *FD*, i. 127–36), which commemorates a Burgundian victory over French troops at Guinegatte in 1479. *j*'s compiler follows this piece with a 'Response aud' canticque faicte et composee en la ville de tournay ꝑ le desßd' johannes nicolay' (fo. 348ᵛ), which refutes the *Journee*'s version of events.[79] Hence a combination of rubrication and *mise en recueil* presents Molinet as a blinkered Burgundian partisan.

Other manuscripts which foreground the documentary value of Molinet's poetry in this way include BR, ms. II, 2545; this consists of Jean de Haynin's *Mémoires*, into which various verse texts have been inserted, including Molinet's *Temple de Mars* (fos. 275ʳ–280ᵛ; *FD*, i. 65–76). The inserted poems are always related to historical events, in the *Temple*'s case the Treaty of Soleuvre of 1475. They consequently illustrate Jean's account; indeed, in the latter part of the volume, where the *Temple* appears, the accumulation of documentary materials seems to take precedence over Jean's narrative of events.[80] Similarly, an epitaph for Charles the Bold, very probably by Molinet, appears in two manuscripts of Louis de La Fontaine's *Histoire de Valenciennes*, completed in 1553.[81] These and other volumes stress the extent to which much of Molinet's poetry

[77] See Dupire, *Étude*, 70–1.

[78] On the *Bergier* and the Peace of Arras, see Devaux, *Jean Molinet*, 433–4, 517–19. For *j*'s critical introductory rubric to Molinet's *Resource du petit peuple*, see Armstrong, 'Practice', 280.

[79] Nicolay's piece also follows the *Journee* in BNF ms. fr. 2375, fos. 77ʳ–80ᵛ; see Dupire, *Étude*, 60.

[80] On the Treaty, see Molinet, *Chroniques*, i. 115–27; Devaux, *Jean Molinet*, 550–3. Dupire, *Étude*, 77, and A. Bayot, 'Notice du ms. original des *Mémoires* de Jean de Haynin', discuss the manuscript; A. C. de Nève de Roden, 'Les *Mémoires* de Jean de Haynin: Des "mémoires", un *livre*', examines Jean's writerly practice.

[81] See H. Servant, 'Un poème inédit de Molinet à la Bibliothèque Municipale de Valenciennes'.

does not simply exhibit the author's ingenuity, but constitutes reactions to current and recent events, informed by loyalty to the Austro-Burgundian dynasty, confidence in chivalric values, and a desire for peace.[82]

Conclusions

As the previous section shows, the self-conscious aspects of Molinet's poetry are sometimes relegated to a secondary position behind its historical or didactic concerns. However, manuscripts with these priorities do not always neglect Molinet's work as writer. In *e*, the frequency of attribution draws attention to Molinet's authorship, while the heading for a section of the *Resource du petit peuple* highlights his use of the *balade fatrisée* form to express Justice's final pleas for peace: 'Lad' iustice parle en mutation de dictz' (fo. 78^{r}; *FD*, i. 159–60). Such volumes, then, present the poetry's self-conscious features as complementing the political discourse, not as submerged by it. Manuscripts which simply ignore Molinet's writerly investment in his poetry are relatively few in number, and often contain little of his material: the *Epitaphe du duc Philippe de Bourgogne* (*FD*, i. 34–5), for example, appears in numerous manuscripts, often devoted to historical material relating to Burgundy, but is frequently the only Molinet piece in these volumes.[83]

Clearly, scribes frequently respond expressively to Molinet's rhetoric, rightly implying that prosodic and linguistic anfractuosities are characteristic and important in his work. Many presentational techniques, such as displayed rhymes and the variation of script in bilingual texts, have precedents in contemporary manuscripts; however, their concentration in the manuscripts discussed suggests that scribes are not simply reverting to standard practices, but reacting to the rhetorical features of individual poems in many cases. Nor is layout merely an aid to readerly comprehension: this is true in some cases, but such techniques as the pictorial representation of rhyme-words make the reader's task more difficult. The page is a site upon which the scribe's virtuosity can comple-

[82] Devaux, *Jean Molinet*, 593–9, summarizes the ideological basis of Molinet's writing.

[83] On the *Epitaphe*'s transmission, see A. Armstrong, '*Grand Rhétoriqueur* Poetics: From Manuscript to Print', D.Phil. thesis (Oxford, 1995), 78–81.

ment and rival that of Molinet. Other self-conscious techniques, less common in Molinet's verse, are sporadically highlighted. Thematization of the poet's activity; reflexivity in correspondence pieces; the knowing play of parodic poems upon an intertext; allusions to Molinet as author: all are brought to readers' attention in different manuscripts.

It is not only through the foregrounding of first-person narration and the like that Molinet's extratextual life as author is suggested to the reader. The existence of anthologies devoted solely to Molinet pieces, and crediting his authorship, implies a measure of interest in the poet's identity, rather than purely in the form or content of his work. Other manuscripts also sometimes attribute Molinet's poems, thereby acknowledging his activity as poet. As might be expected, however, general tendencies as to texts' attribution cannot easily be identified. Molinet is less likely to be credited in earlier manuscripts,[84] while his authorship is acknowledged more often in later volumes, such as *e* or the Cambridge manuscript, perhaps under the influence of print. The latter tendency, however, is inconsistent; Mons, Archives de l'État, ms. divers 114, does not credit Molinet, and BR, ms. IV, 541, dated 1568, does not always do so.[85] Similarly, Molinet's work is attributed more frequently in manuscripts which emphasize his poetry's self-conscious aspects than in those which privilege its historical content: the *Epitaphe du duc Philippe de Bourgogne* is anonymous in all but one of the manuscripts in which it appears.[86] Yet *e*, with its historical concerns, attributes every Molinet piece it contains, while *h*, where formal artistry is prominent, bears no attributions.

The difficulty of achieving a clear overview of manuscript attributions is perhaps hardly surprising; the wide range of

[84] The following 15th-century manuscripts containing Molinet poems bear no explicit attributions to Molinet: Arsenal, ms. 3521; BR, ms. II, 2545; BNF, ms. fr. 1642 (see Dupire, *Étude*, 51); BR, ms. 21521–21531 (see ibid. 75–6); BR, ms. IV, 144 (see Bibliothèque Royale Albert Ier (Brussels), *Quinze années d'acquisition: De la pose de la première pierre à l'inauguration officielle de la Bibliothèque* (Brussels, 1969), 108); Arnhem, Bibliotheek, MS 79, which contains the *Naufrage* (fos. 60^r–69^r; see A. Chartier, *Poetical Works*, ed. J. C. Laidlaw (Cambridge, 1974), 124–5, and a *notice* in the IRHT); BNF, n. a. fr. 1541 and Saint-Omer, BM, ms. 657, which both contain the *Temple de Mars* (fos. 48^r–54^r, 108^r–113^r; see respectively H. Omont, *Bibliothèque Nationale. Catalogue général des manuscrits français: Nouvelles Acquisitions françaises* (Paris, 1899–1918), i. 233, and a *notice* in the IRHT).

[85] On BR, ms. IV, 541, see J. Lemaire, 'Meschinot, Molinet, Villon: Témoignages inédits. Étude du Bruxellensis IV 541, suivie de l'édition de quelques ballades'.

[86] Armstrong, '*Grand Rhétoriqueur* Poetics', 79.

Molinet's output permits the compilation of manuscripts with very different priorities, according to the poems selected. Manuscripts focusing on historical affairs from a documentary viewpoint, politically oriented volumes exploiting Molinet's pro-Burgundian rhetoric, and books valorizing poetic craft exist alongside the major Molinet anthologies with their more eclectic preoccupations. In the light of this variation, it is all the more noteworthy that so many scribes acknowledge, if not Molinet's authorship, then at least the formal self-consciousness of his poetry.

2

EDITIONS OF MOLINET'S POETRY

Introduction

In *La Recollection des merveilleuses advenues*, Molinet acknowledges printing's educational benefits:

> J'ay veu grant multitude
> De livres empraintés,
> Pour tirer en estude
> Povres mal argentés;
> Par ces nouvelles modes
> Auront maintz escolliers
> Decrets, bibles et codes,
> Sans grand argent baillier. (*FD*, i. 307, vv. 553–60)

Yet it is uncertain how far Molinet exploited the medium to his own advantage. Not until over twenty years after his death was a substantial collection of his poetry published; this anthology, *Les Faictz et dictz*, appeared three times in Paris between 1531 and 1540. Editions of single Molinet poems are quite numerous, but are mostly unattributed, suggesting that they were published without Molinet's involvement.[1] However, Jean de Liège printed three editions during Molinet's lifetime in Valenciennes; since Molinet lived in this town, it has been considered likely that he co-operated in these editions' publication.[2]

Most printed versions of Molinet's poetry, then, must be regarded as indices of reception, not projects sanctioned by the poet; in this respect, they resemble manuscript versions. In the

[1] Dupire, *Étude*, 87–112, lists both collective and single-text editions of Molinet's poetry, as well as multi-author collections which contain Molinet pieces. To his listing should be added the various collections containing the *Complainte d'un gentilhomme a sa dame* (see F. Lachèvre, *Les Recueils collectifs de poésies libres et satiriques publiés depuis 1600 jusqu'à la mort de Théophile (1626)* (Paris, 1914), 181), and a ninth edition of the *Temple de Mars* (Paris, no pub., n.d. [Alain Lotrian, *c.*1531]; Paris, Bibl. de l'École Nationale Supérieure des Beaux-Arts, Rés. Masson 469).

[2] See R. Giard and H. Lemaître, 'Les Origines de l'imprimerie à Valenciennes: Jehan de Liège'; G. Colin, 'Jean de Liège'; H. Servant, 'Jehan de Liège, premier imprimeur valenciennois'.

absence of authorial influence, both scribes and printers have a relative freedom in dealing with a text, though their own horizon of expectations always constrains them. Where printers differ from scribes is in their more immediate subordination to market forces; the need to appeal to a wide and less knowledgeable audience may cause printers' literary priorities to differ from those of scribes and owners in a manuscript-centred milieu. Consequently, we might expect the two media to present the self-conscious aspects of Molinet's poetry in different ways. My study of the editions of Molinet's poems focuses upon the contrasts which they present with both the general tendencies, and the presentation of individual texts, in manuscripts.

The obvious starting-point for this examination is the *Faictz et dictz*, a collection comparable to the major manuscript anthologies, and broadly contemporary with several later manuscripts of Molinet's verse. I then analyse two groups of single-text editions which merit especial attention: the Valenciennes editions, for obvious reasons; and editions of the *Temple de Mars*, Molinet's most frequently printed piece, whose presentation is broadly representative of that of single-text editions. Comparisons are occasionally adduced to editions of other pieces, and to printed multi-author collections containing Molinet poems. In conclusion, I assess the presentational differences between manuscripts and editions of Molinet's poetry, and address the question of Molinet's participation in the publication of the Valenciennes editions.

Les Faictz et dictz

Les Faictz et dictz was first published in 1531, by Jean Longis and the widow of Jean Saint-Denis, with characters and ornaments from the printer Nicolas Couteau: Longis must be regarded as the principal publisher, since the edition bore a privilege granted to him.[3] Subsequent editions, which appeared in 1537 and 1540, had six and five printers respectively. Each edition's collaborative pub-

[3] BNF, Rés. Ye 41, fo. ♣1^{v}. Further references to the printed *Faictz* are to this edition, unless the use of another edition is made clear by the context, and will be incorporated into the text. On this edition's printing, see B. Moreau, *Inventaire chronologique des éditions parisiennes du XVIe siècle* (Paris, 1972–), iv. 244. Dupire, *Étude*, 87–98, provides details of all the editions, including the various printers' addresses on title-pages and a listing of the first edition's contents.

lication, presumably to permit costs to be shared, implies that producing the *Faictz* was a considerable enterprise, even for such well-established Paris printers as Longis. The financial stakes involved are indicated by the first edition's prestigious folio format and unusually generous privilege, granted for four years rather than the usual two or three.[4] Clearly, the publication of Molinet's poetry was regarded as a sound commercial venture, worth protecting from unauthorized reprints, and likely to provide a return on printers' investments. The large sums obviously spent on publication also imply that economic constraints are less likely to have influenced the poems' visual presentation than in more cheaply produced volumes, where space is at a premium.

The successive editions exhibit significant physical differences, reflecting the printed book's evolution in the early sixteenth century. While the first edition is a folio volume in bastard type, that of 1537 adopts an octavo format, and that of 1540 introduces roman type. However, the later editions clearly derive from the first. Each edition contains the same sequence of texts, including an anonymous preface probably written by the collection's first editor, which praises Molinet's work and draws attention to the processes of compilation and publication. My analysis focuses, therefore, upon the 1531 edition, and addresses the later editions only insofar as the changes in their presentation significantly modify the effects of their predecessor's material form.

Molinet's name is prominent in the 1531 *Faictz*, both on the title-page (Plate 2.1) and in the preface; his identity in itself was apparently of interest to purchasers. As in the major manuscript anthologies, a wide range of poems coheres around a single author, showing his range of talents to advantage. Indeed, the editor's preface mentions 'la variete des matieres' as a reason to 'affectueusement appeter et desirer' the anthology (fo. ♣3^{v}). Yet further examination reveals that both the preface and the anthology's contents promote a more particular reading of Molinet's work than the editor suggests.

[4] Longis had published only three folio editions by 1531, all chronicles or prose romances shared with other publishers: see Moreau, *Inventaire*, iii. 1413, 1428, 2239. His editions of poetry were usually in octavo throughout his career. On the duration of privileges, see E. Armstrong, *Before Copyright: The French Book-Privilege System 1498–1526* (Cambridge, 1990), 118–25.

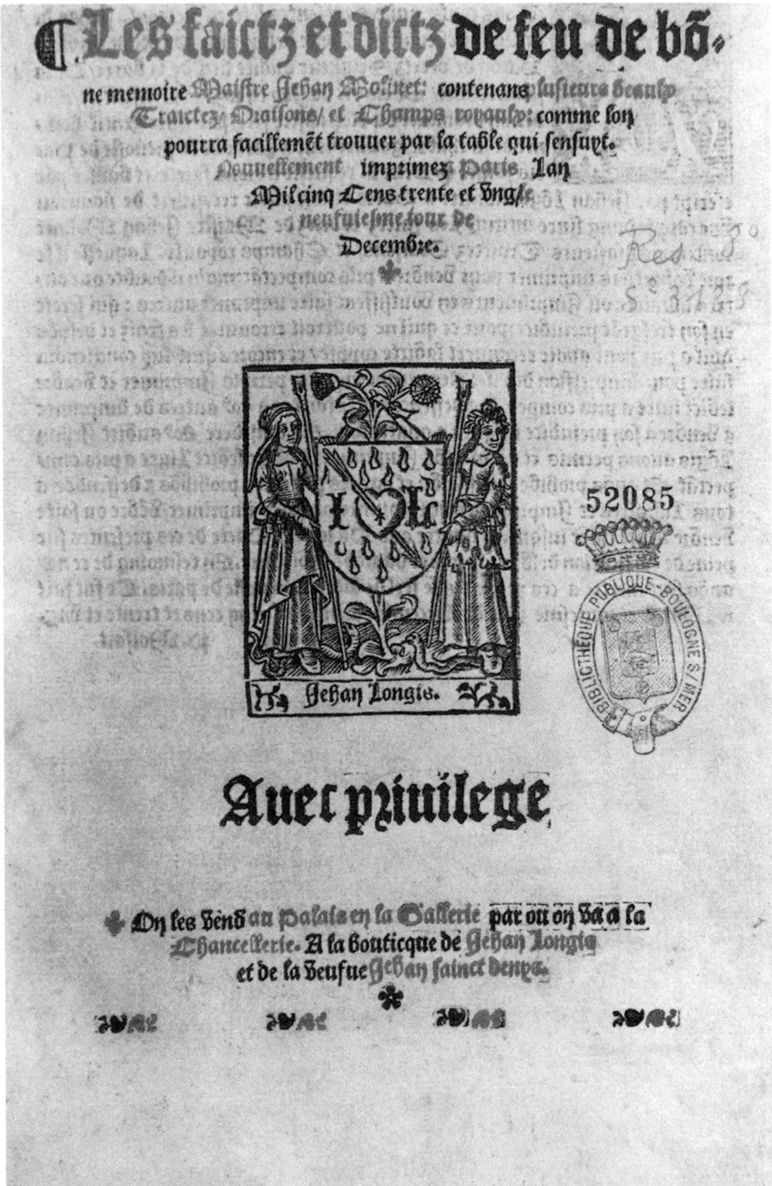

C Les faictz et dictz de feu de bō-
ne memoire Maistre Jehan Molinet: contenans plusieurs beaulx
Traictez, Oraisons, et Champs royaulx: comme lon
pourra facillemēt trouuer par la table qui sensuyt.
Nouuellement imprimez Paris. Lan
Mil cinq Cens trente et vng/le
neufuiesme iour de
Decembre.

Jehan Longis.

Auec priuilege

On les vend au Palais en la Gallerie par ou on va a la
Chancellerie. A la bouticque de Jehan Longis
et de la veufue Jehan sainct denys.

PLATE 2.1 Title-page of the 1531 *Faictz et dictz* (Boulogne-sur-mer, Bibliothèque Municipale)

Editorial prefaces may have an informative or commendatory role, or both.[5] In this case, praise outweighs analysis:

Se sont nagueres presentez ⁊ mis au deuãt aucuns singuliers traictez cõposez par feu maistre Iehan molinet tresexpert en lart et science de rhetoricque francoise/ lesquelz sont si abundans en termes exquis illustrez de grans sentẽces et dictz moraulx non moins fructueux ⁊ vtiles q̃ ioyeux et delectables a tous lecteurs et auditeurs ayans ledit art ⁊ science en amour ⁊ affection/ quon pourroit par ce facillemẽt iuger icelluy autheur auoir souuẽt frequẽte auec les muses es mõs parnasus ⁊ helicon ⁊ illec este alimẽte ⁊ suffisammẽt abreuue de la doulce ⁊ melliflue liqueur procedãt de la fontaine caballine. Affin dõcques que le dit autheur ne soit du tout frustre du loz quil a iustement merite doubtant que par laps et diuturnite de temps sesdictes oeuures expirẽt et deperissent. Pour iceulx reuoquer des vmbres ⁊ tenebres de obliuion et reduire en vraye clarte et lumiere tellement que chascun en puisse auoir certaine iouyssance ont este lesdictes oeuures espars en diuers lieux recueillis et ensemble congerez et compile ce present petit volume. (fo. ♣ 3^{v})

No biographical details are provided, in contrast to many contemporary editorial prefaces, which provide at least some information about the author. The *Recueil Jehan Marot*'s preface, for instance, states that the poet knew no Greek or Latin.[6] It is perhaps due to the editor's apparent ignorance of Molinet's career that the preface also includes no information on his Burgundian loyalties. However, this may be a matter of editorial policy; as I shall demonstrate, the volume's selection of poems, and certain textual variants, attenuate Molinet's more anti-French attitudes.

The preface assesses Molinet's work in two principal ways. First, moral genres and attitudes are privileged. 'Traictez', for instance, suggests discursive and utilitarian texts;[7] indeed, the term appears on the title-page alongside the devotional forms 'Oraisons' and 'Champs royaulx'. 'Sentences' and 'dictz moraulx' also suggest that the interest of Molinet's poetry lies in its didactic utility. This evaluation is probably at least partly formulaic: the editor's later claim that 'pourront tous beniuoles euitant oysiuete mere de tous vices en icelluy prendre quelque fruict et recreation' (fo. ♣ 3^{v}) combines the venerable topos of utility and pleasure with the equally

[5] See A. Ruppen, 'Bacchus René: Du "livre" au "moi" dans quelques préfaces de traducteurs à la Renaissance', 21.

[6] J. Marot, *Recueil*, ed. A. Saunders (Menston, 1971), fo. $A2^{r}$.

[7] *Traitié*, 'ouvrage où l'on traite d'une matière, en en examinant par ordre les diverses parties': F. Godefroy, *Dictionnaire de l'ancienne langue française* (Paris, 1881–1902), x. 797.

traditional justification of reading as a defence against idleness.[8] Other contemporary editorial prefaces often assert more thoroughly the ethical value of the texts they introduce: Galliot Du Pré's *Preambule* to his edition of Alain Chartier's works cites Aristotle and Plutarch as examples of moralizing literature which justify his publication of Chartier's 'matieres tresutilles'.[9] Nevertheless, the preface's moral focus is undeniable, especially since the title-page echoes its concerns. Molinet's eloquence is seen largely as serving didactic ends, an evaluation which homogenizes his diverse output and denies the importance of both his specifically political works and his parodic and bawdy pieces.

Secondly, the preface presents Molinet as 'tresexpert en lart et science de rhetoricque'. This might seem almost otiose, given Molinet's reputation as a rhetorician, acknowledged by Fabri and others,[10] but close comparison of Molinet's practice with the preface reveals that the editor and Molinet understand 'rhetoricque' differently. As I have previously outlined, both Molinet's poetry and his *Art de rhétorique* evince a view of rhetoric as principally consisting of elaborate versification and linguistic virtuosity. The editor's reference to rhetoric as an 'art et science' apparently echoes this conception, implying that rhetoric depends on formal rules which may be learned. Yet the preface's image of the Muses suggests that poetry is conceived as primarily proceeding from inspiration. While this motif was not unknown to the *rhétoriqueurs*,[11] it is not so central to *rhétoriqueur* poetics as the valorization of technical difficulty; it is exploited more by later generations, particularly the Pléiade.[12] Once again, the preface assesses Molinet in terms other than those of his own practice, reflecting the ways in which tastes in the literary market place had changed in the generation after his death. The aspects of literary self-consciousness characteristic of Molinet's poetry, particularly prosodic intricacy, are not acknowledged.

[8] See E. R. Curtius, 'Der Archipoeta und der Stil mittellateinischer Dichtung', 136.

[9] A. Chartier, *Les Faictz et dictz*, Paris, Galliot Du Pré, 1526 (Bodleian Library, MS E. D. Clarke 33), fos. a2^{v}–a3^{r}. Winn, *Anthoine Vérard*, 57–8, discusses the didactic elements in Vérard's prologues.

[10] See Fabri, *Grand et vrai art*, ii. 61–2.

[11] See Zsuppán, 'Early Example'.

[12] See G. Castor, *Pléiade Poetics: A Study in Sixteenth-Century Thought and Terminology* (Cambridge, 1964), chs. 3–4.

The preface also valorizes the editor's activity, as was common practice in the period.[13] It is the editor's achievement in compiling and publishing the collection, rather than the poet's talent, which is presented as securing 'loz' and 'clarte' for Molinet's work. Indeed, the editor almost usurps Molinet's role, through his use of mythology. His references to the Muses, Parnassus, and Helicon echo the *rhétoriqueur* technique of *poetrie*, the use of mythological fictions with an allegorical sense.[14] By thus appropriating a distinctively poetic mode of expression, the editor makes readers aware of his own literary talents. Hence the preface reveals a high degree of editorial self-consciousness, despite its lack of first-person discourse on the editor's part. Printing's economic imperatives make Molinet's poetry into a pretext for the editor's profit-oriented discourse: poems are praised so that books will be sold. The editor promotes himself at the expense of the poetry's own self-conscious features: the preface's accentuation of the texts' didactic value is a case in point.

The contents and their order confirm the preface's apparent moralizing bias. The collection is dominated by didactic, religious, and historical material, including all Molinet's *prosimetrum* pieces and many prayers. Some correspondence pieces and poems employing unusual formal devices appear, but carry little combined weight in the anthology as a whole.[15] Similarly, while poems like the *Dyalogue du gendarme et de l'amoureux* (*FD*, ii. 616–27; fos. Q5^{v}–R1^{v}) contain sexual double entendres, there are no explicitly bawdy pieces. This selectivity, whether or not dictated by available sources, contrasts with the wider range of the principal manuscript anthologies. Moreover, Molinet's pro-Burgundian attitudes are moderated. Not only does the *Faictz* lack poems devoted

[13] Contemporary translators' prefaces often stress the translators' own efforts and merits above those of the original author and his work: see Ruppen, 'Bacchus René'. Similarly, Clément Marot's preface to his edition of Villon's works mentions his emendations of corrupt editions, highlighting his philological expertise. This preface is edited in B. Weinberg, *Critical Prefaces of the French Renaissance* (Evanston, Ill., 1950), 65–8. For analysis of Marot's editorial practice, see M. Lazard, 'Clément Marot éditeur et lecteur de Villon'; C. Dop-Miller, 'Clément Marot et l'édition humaniste des œuvres de François Villon'. On the self-promotional strategies in Anthoine Vérard's dedicatory prologues, see Winn, *Anthoine Vérard*, 52–65.

[14] See M.-R. Jung, 'Poetria: Zur Dichtungstheorie des ausgehenden Mittelalters in Frankreich'.

[15] e.g. *Ung present fait a l'empereur*, 'Domine mi reverende' (fos. V4^{r}, V6^{v}).

to Burgundian victories over France, but certain poems display variants which reduce their anti-French orientation. The *Recollection des merveilleuses advenues* omits stanzas alluding to the French defeat at Guinegatte and the recapture of Arras from France in 1492,[16] while variant lines in the *Temple de Mars* promulgate a Franco-imperial alliance to defend Christendom.[17] The poems' order reinforces the moralizing trend. A series of prayers, and other religious and didactic poems, opens the collection; occasional and didactic pieces complete the first half of the book (fos. A1^{r}–M4^{v}). The second half contains a mixture of occasional pieces and lighter material; most of the final poems are moralizing and historical texts. Lighter verses, correspondence pieces, and poems with unusual layouts appear largely in the third quarter of the volume, with didactic texts in the more prominent opening position. While many religious, moralizing, and occasional pieces employ highly sophisticated rhetoric and versification, this ordering, in conjunction with the preface, foregrounds their historical and didactic content.

Layout is generally clear. Most texts, and their major subdivisions, are clearly marked by woodcut initials; titles are frequent and clearly spaced; *explicits* conclude over half the poems. Consequently, poems are visually discrete. Moreover, stanzas are marked with small type ornaments (paraphs, fleurs-de-lis, or *fleurons*), which alternate from one stanza to the next, rendering them more distinct on the page. This is not solely a matter of house style: type ornaments rarely alternate in Longis's other editions of poetry, such as Cretin's *Le Debat de deux dames*,[18] while in *Les Motz dorez de Cathon* they alternate between titles and texts, not between indi-

[16] *FD*, i. 284–334, stanzas 84, 130.

[17] Most versions contain the lines:

> Ne resemblés Olopherne, Alixandre,
> Felle calandre, ostrice ou salemandre
> Qui feu de flandre et fer veullent mengier. (vv. 291–3)

In the *Faictz*, however, this passage is replaced by:

> Ne resemblez holoferne alexandre
> Vueillez entendre auec la sallemande [*sic*]
> Par cueur de flandre contre turcz rereuanger [*sic*]. (fo. L4^{r})

See *Recueil de pièces historiques imprimées sous le règne de Louis XI*, ed. É. Picot and H. Stein (Paris, 1923), 17.

[18] Paris, Antoine Couteau for Jean Longis, 1 April 1526 (BNF, Rés. Ye 1337).

vidual stanzas.[19] The clear grammar of legibility apparently responds to Molinet's verse on its own terms.

Most significantly, there is a change in layout halfway through the *Faictz*. In earlier pieces, verse is usually presented in a single column; only a few stanzas are in two columns. However, from *La Complainte de Renommee* (*FD*, i. 193–204; fos. I5^{v}–K1^{v}), a two-column layout becomes the norm for verse, though prose remains in long lines. The few exceptions are mostly pieces with unconventional layouts, such as *Ung present fait a l'empereur*. Two further developments approximately coincide with this change. From fo. I5^{r}, larger woodcut initials tend to be used, occupying six lines, rather than three or four as in most earlier pieces; from gathering L, a large textura typeface, otherwise used only on the title-page, is used for the first line of many titles. The page's visual diversity, and the area of decorative space relative to text, are thus increased.

These changes may be due to workshop practice, to economic considerations (the saving of paper), or to the manuscript source; but whatever the reason for it, the new layout makes new demands upon readers. By retaining long lines for prose, the double-column layout forces readers to adopt two different reading habits. With prose, the eye can move down the page without needing to move back up unless the reader so wishes, as in a normal single-column layout;[20] with two-column verse, the eye must move back up the page after finishing the first column. These contrasting physical demands accentuate the difference between verse and prose, making readers more aware of the ways in which Molinet combines the two forms.

This effect is not inherent to the two-column verse layout, frequently used in contemporary printed books whose prose was in long lines. However, readers of this edition are already accustomed to a single-column layout when the use of columns abruptly changes, requiring them to modify their habits. The shift has specific implications because it occurs within a single book, and because the dense formal patterns in Molinet's poetry often

[19] Paris, no pub., n.d. [Guillaume de Bossozel and Jean Longis, *c.*1530] (BNF, Rés. p. Yc. 906).

[20] This does not, however, imply that reading is an exclusively linear process: see K. S. Goodman, *Language and Literacy: The Selected Writings of Kenneth S. Goodman*, ed. F. V. Gollasch (London, 1982).

invite significant visual presentation. Indeed, the change in the use of initials, typefaces, and columns may be seen as marking a slightly different aesthetic. The single-column layout, normal in both manuscript and print, enables verse and prose to be read in relatively linear fashion. The two-column layout, in this context, draws more attention to versification. Prose and verse are more clearly differentiated, while the retention of a single column for pieces with unusual layouts cannot but accentuate their expressive use of the page. Moreover, the introduction of a different typeface for titles marks a more sophisticated grammar of legibility, while the larger initials indicate poems' beginnings or internal articulations more clearly. All these effects suggest that the second layout draws more attention to Molinet's versification. It is perhaps no coincidence that the texts which make most expressive use of the page appear in the second part of the *Faictz*.

The contrast between the layouts should not be overstated: the differences are only relative. Nor are the layouts wholly exclusive: two-column verse appears amid a one-column layout and vice versa, varying the page's appearance and underlining the variety of media Molinet uses. Nevertheless, the second layout is visually more elaborate, places greater emphasis on formal features, and can produce subtle effects. A telling instance of such effects appears in *Le Trosne d'Honneur* (fos. F5^{r}–G5^{r}), which recounts the ascent of a flower, metaphorically denoting the recently deceased Philip the Good, through nine heavens to sit at Honneur's right hand.[21] On reaching the third heaven, Philip is greeted by 'Monseigneur Cornille, bastard de Bourgongne et messire Jacques de Lalaing' (*FD*, i. 49, ll. 49–50), who greet him in a *double fatras*. In the *Faictz*, the two stanzas of this poem are presented in parallel columns (fo. G3^{r}; Plate 2.2), a layout one is tempted to read as denoting simultaneity, as if each speaker delivers one stanza at the same time in a polyphonic address.[22]

The *Faictz* also includes some poems which incorporate acrostics and rebuses, with layouts which respect their visual potential.

[21] See S. S. Sutch, 'Comment lire une allégorie de personnification séculière'.

[22] Contemporary musical manuscripts do not represent polyphony in this way, but generally present different parts of a song sequentially: see W. Apel, *The Notation of Polyphonic Music 900–1600* (5th edn. Cambridge, Mass., 1953), 103–9. Hence this volume offers more visual simultaneity than is usual in manuscripts of music. Similar layouts appear in two manuscripts of the *Trosne*, BNF, n. a. fr. 21532 (fo. 7^{v}) and Courtrai, Stedelijke Openbare Bibliotheek, ms. 366 (fos. 6^{v}–7^{r}).

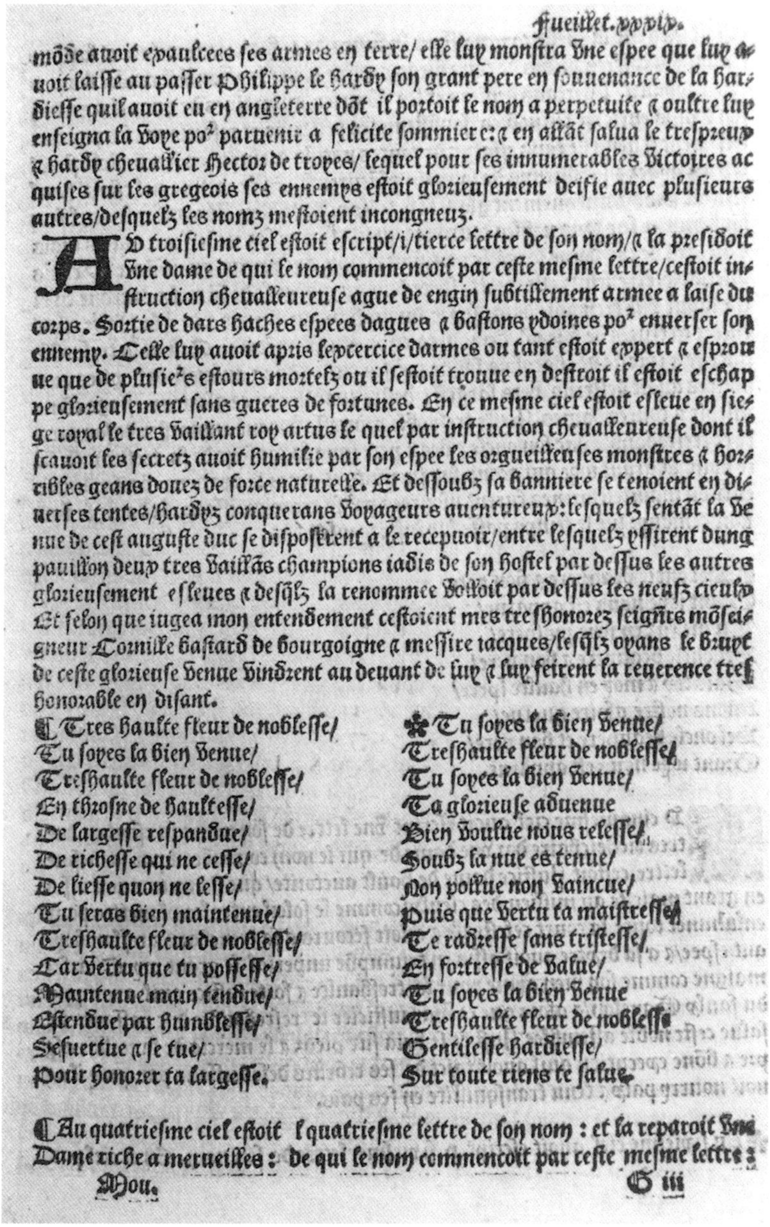

ffueillet.xxxix.

mõde auoit espaulcees ses armes en terre/ elle luy monstra vne espee que luy a-
uoit laisse au passer Philippe le hardy son grant pere en souuenance de la har-
diesse quil auoit eu en angleterre dõt il portoit le nom a perpetuite & oultre luy
enseigna la voye po[2] paruenir a felicite sommiere: & en allãt salua le trespreux
& hardy cheuallier Hector de troyes/ lequel pour ses innumerables victoires ac
quises sur les gregeois ses ennemys estoit glorieusement deifie auec plusieurs
autres/desquelz les nomz mestoient incongneuz.

AU troisiesme ciel estoit escript/i/tierce lettre de son nom/& la presidoit
vne dame de qui le nom commencoit par ceste mesme lettre/cestoit in-
struction cheualleureuse ague de engin subtillement armee a laise du
corps. Sortie de dars haches espees dagues & bastons ydoines po[2] enuerser son
ennemy. Celle luy auoit apris lexercice darmes ou tant estoit expert & esprou
ue que de plusie[2]s estours mortelz ou il sestoit trouue en destroit il estoit eschap
pe glorieusement sans gueres de fortunes. En ce mesme ciel estoit esleue en sie-
ge royal le tres vaillant roy artus le quel par instruction cheualleureuse dont il
scauoit les secretz auoit humilie par son espee les orgueilleuses monstres & hor-
ribles geans douez de force naturelle. Et dessoubz sa banniere se tenoient en di-
uerses tentes/hardys conquerans voyageurs auentureux: lesquelz sentãt la ve-
nue de cest auguste duc se disposerent a le recepuoir/entre lesquelz yssirent dung
pauillon deux tres vaillãs champions iadis de son hostel par dessus les autres
glorieusement esleues & desqlz la renommee volloit par dessus les neufz cieulx
Et selon que iugea mon entendement cestoient mes treshonorez seigñrs mõsei-
gneur Cornille bastard de bourgoigne & messire iacques/lesqlz oyans le bruyt
de ceste glorieuse venue vindrent au deuant de luy & luy feirent la reuerence tres
honorable en disant.

¶Tres haulte fleur de noblesse/
Tu soyes la bien venue/
Treshaulte fleur de noblesse/
En throsne de haultesse/
De largesse respandue/
De richesse qui ne cesse/
De liesse quon ne lesse/
Tu seras bien maintenue/
Treshaulte fleur de noblesse/
Car vertu que tu possesse/
Maintenue main tendue/
Estendue par humblesse/
Se suertue & se tue/
Pour honorer ta largesse.

Tu soyes la bien venue/
Treshaulte fleur de noblesse/
Tu soyes la bien venue/
Ta glorieuse aduenue
Bien voulue nous relesse/
Soubz la nue es tenue/
Non pollue non vaincue/
Puis que vertu ta maistresse
Te radresse sans tristesse/
En fortresse de salue/
Tu soyes la bien venue
Treshaulte fleur de noblesse
Gentilesse hardiesse/
Sur toute riens te salue.

¶Au quatriesme ciel estoit l quatriesme lettre de son nom : et la reparoit vne
Dame riche a merueilles : de qui le nom commencoit par ceste mesme lettre :

Mon. G iii

PLATE 2.2 Parallel columns for the *double fatras* in the *Trosne d'Honneur*, 1531 *Faictz*, fo. G3^r (Boulogne-sur-mer, Bibliothèque Municipale)

Yet these poems' presentation is generally less expressive than in manuscripts. Latin and French, for instance, are not visually differentiated in bilingual texts; the greatest differentiation occurs in *A maistre David Walle* (fo. S2^{v}), where the French verses are indented. Four pieces have displayed rhymes; three others, which have displayed rhymes in various manuscripts, do not.[23] In general, layouts are less varied and expressive than those of many contemporary printed books. Anthoine Vérard's *Louenges a Nostre Seigneur* anthology prints the Latin text incorporated in Lemaire's bilingual *Salve Regina* in red, and uses spacing to highlight the letters forming an acrostic.[24] Editions of André de La Vigne's *Ressource de la Chrestienté* also draw attention to acrostics through spacing, though this technique becomes rarer in later editions.[25] Printing in red, and explanatory annotations, foreground the use of letters to represent rulers and states in editions of other political poems by La Vigne.[26] Latin and French were often printed in different typefaces, as in Longis's own *Motz dorez de Cathon*. Indeed, Longis's other editions of poetry frequently exploit texts' visual potential more thoroughly than does the *Faictz*. Verse is usually laid out conventionally, but is not formally intricate enough to invite iconically expressive layouts, unlike much of Molinet's work. Significantly, Longis invests in woodcuts where appropriate, even early in his career and in octavo editions: Cretin's *Debat de deux dames* includes three cuts, representing the birds and hunting-dogs whose relative merits are debated. In sharp contrast, the *Faictz* does not reflect the considerable visual potential of Molinet's allegories through illustration. Neither can this relative plainness be attributed to economic constraints: as previously noted, the collection's presentation is expensive. The relative neglect of poems' visual potential is not due to technical or financial considerations,

[23] Displayed rhymes are used for *Magistro Ludovico Compere* (*FD*, ii. 779), *Ung present fait a l'empereur*, *Lettre a Monseigneur l'archiduc quand il alla en Espaine* (*FD*, i. 371–2), and 'Domine mi reverende' (fos. V4^{r}, V6^{r-v}). They are absent from *Lettre a maistre Anthoine Busnois* (*FD*, ii. 795–6), 'Manne du ciel, doulce fleur de concorde' (*FD*, i. 345–6), and *Ballade adreschant a messeigneurs de Foix, Monpensier et Vendosme* (*FD*, i. 392–3) (fos. S2^{r}, V3^{r-v}, Z4^{v}–5^{r}). Of the latter three pieces, all have displayed rhymes in manuscript *C*, the first two in *B*, and the second in the Cambridge manuscript.

[24] Paris, Anthoine Vérard, n.d. [*c*.1502] (BNF, Rés. Ye 831), fos. mm2^{r}–3^{v}, ss3^{r}; Winn, *Anthoine Vérard*, 427–31, edits and discusses the latter piece.

[25] A. de La Vigne, *La Ressource de la Chrestienté*, ed. C. J. Brown (Montreal, 1989), 83–9.

[26] C. J. Brown, *The Shaping of History and Poetry in Late Medieval France: Propaganda and Artistic Expression in the Works of the Rhétoriqueurs* (Birmingham, Ala., 1985), app. 3.

but corresponds to the publisher's apparent understanding of rhetoric as evinced in the preface. Though a central aspect of rhetoric for *rhétoriqueurs* and scribes, formal virtuosity is less pre-eminent in the context of 1530s printing; it is not seen as important enough to warrant visual highlighting.

Generally, layouts do readers' interpretative work for them, rather than requiring them to discover formal patterns for themselves; this suggests that the assumed public was not especially well versed in *rhétoriqueur* practice. For the most part, puzzle- and rebus-poems are at least partly solved. The instructions on how to make sense of the verses in the Molinet–Cretin correspondence have already been carried out: in the verse 'Molinet nest sans bruyt ne sans nom non' (fo. V1^{r}; cf. *FD*, ii. 838, v. 1), the third and eighth syllables of the original, 'unsolved' line have been repeated in accordance with the directions 'Doublez le tiers et le huytiesme'. *Ung present fait a l'empereur* is laid out almost as in *C*, with the birds' names which begin and end each line detached to form separate columns, and rhyme-syllables written only once (fo. V4^{r}; Plate 2.3). While this presentation is suggestive, it lacks the challenging quality of *A*'s pictorial representations. In such cases, readers can do little but recognize the poet's ingenuity; they are denied the active discovery of formal patterns.

Clearly, print lends itself to rebus-poems less readily than manuscripts, whose scribes are not required to lock text into space so rigidly. But in 'Domine mi reverende', the 1531 *Faictz* shows that printing's technical constraints do not preclude pictorial rhymes from functioning effectively. Here, as in *C*, small images represent the rhyme-words 'dé' and 'table', and the death's head (fo. V6^{v}; Plate 2.4). This layout is all the more striking for its uniqueness in this edition, though it makes few demands on the compositor. Only three woodcuts are required; the block of verse beside them need not be disturbed, and the images, even if not taken from stock, are generic enough to lend themselves to reuse in later publications.[27] The numerous pictures required for *Ung present* would be not only prohibitively expensive, but also far more difficult to insert in a forme of type. Even in this case, then, the expressiveness of layout is limited. Printing tends to normalize poems whose manuscriptural layouts show elaborate and meaningful use of

[27] Russell, *Emblematic Structures*, 132, suggests that such reusability may have influenced many publishers' decisions.

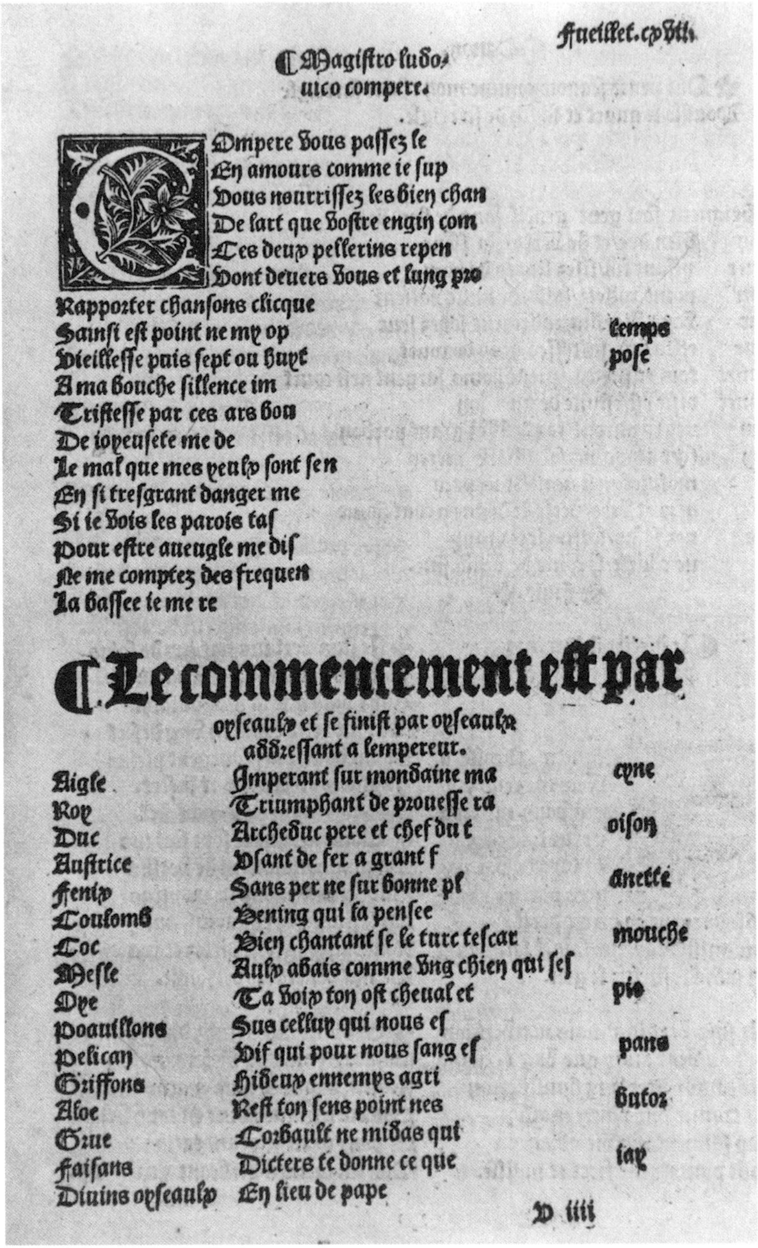

Fueillet.cxviii.

¶ Magistro ludo-
uico compere.

Compere vous passez le
En amours comme ie suy
Vous nourrissez les bien chan
De lart que vostre engin com
Ces deux pellerins repen
Dont devers vous et lung pro
Rapporter chansons clicque
Sainsi est point ne my op — temps
Vieillesse puis sept ou huyt — pose
A ma bouche sillence im
Tristesse par ces ars bou
De ioyeusete me de
Le mal que mes yeulx sont sen
En si tresgrant danger me
Si ie vois les parois tas
Pour estre aveugle me dis
Ne me comptez des frequen
La bassee ie me re

¶ Le commencement est par
oyseaulx et se finist par oyseaulx
addressant a lempereur.

Aigle — Imperant sur mondaine ma — cyne
Roy — Triumphant de prouesse ra
Duc — Archeduc pere et chef du t — oison
Austrice — Usant de fer a grant f
Fenix — Sans per ne sur bonne pl — anette
Coulomb — Bening qui sa pensee
Coc — Bien chantant se le turc tescar — mouche
Mesle — Aulx abais comme ung chien qui ses
Pye — Ta voix ton ost cheval et — pis
Poaulsons — Sus celluy qui nous es
Pelican — Vif qui pour nous sang es — pans
Griffons — Hideux ennemys agri
Aloe — Rest ton sens point nes — butoz
Grue — Corbault ne midas qui
Faisans — Dicters te donne ce que — iay
Divins oyseaulx — En lieu de pape

V iiii

PLATE 2.3 Detached birds' names in *Ung present*; 1531 *Faictz*, fo. V4r (Boulogne-sur-mer, Bibliothèque Municipale)

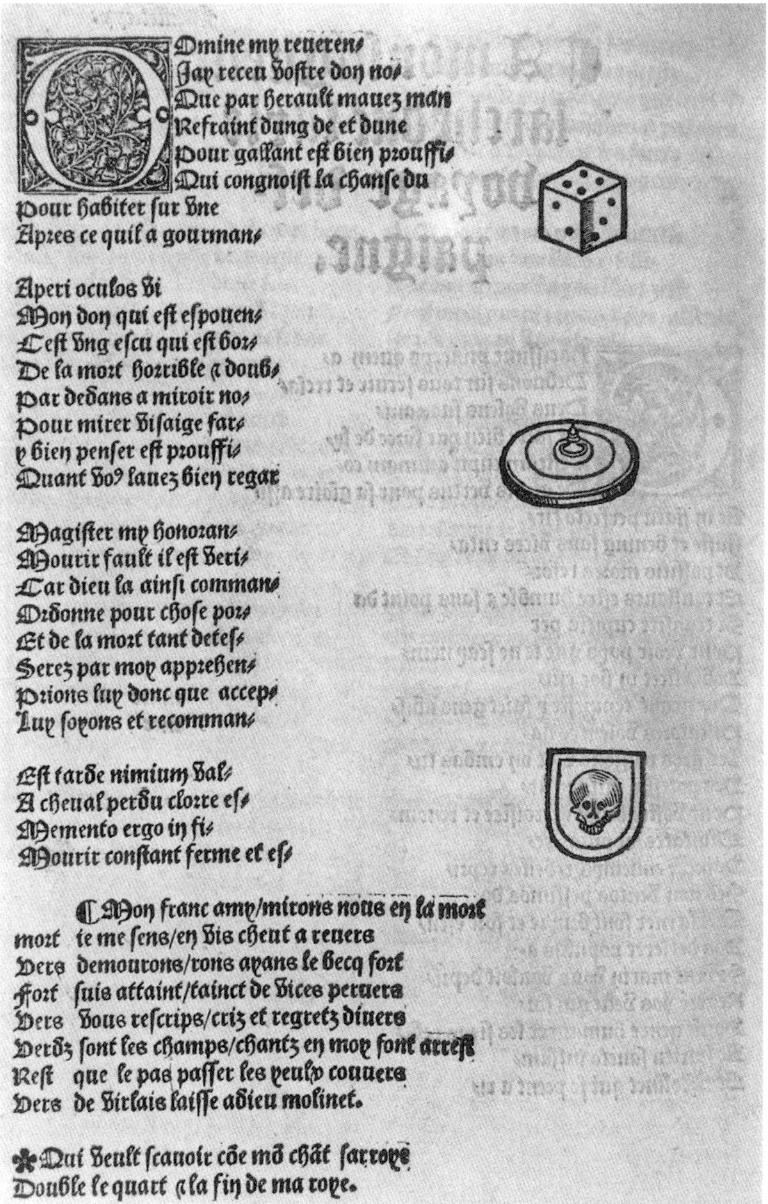

Domine my reueren-
Jay receu vostre don no-
Que par herault mauez man
Refraint dung de et dune
Pour gallant est bien prouffi-
Qui congnoist la chanse du
Pour habiter sur vne
Apres ce quil a gourman-

Aperi oculos vi
Mon don qui est espouen-
Cest vng escu qui est hor-
De la mort horrible & doub-
Par dedans a miroir no-
Pour mirer visaige far-
y bien penser est prouffi-
Quant vo⁹ lauez bien regar

Magister my honoran-
Mourir fault il est veri-
Car dieu la ainsi comman-
Ordonne pour chose por-
Et de la mort tant detes-
Serez par moy apprehen-
Prions luy donc que accep-
Luy soyons et recomman-

Est tarde nimium val-
A cheual perdu clorre es-
Memento ergo in fi-
Mourir constant ferme et es-

¶ Mon franc amy/mirons nous en la mort
mort te me sens/en vis cheut a reuers
Vers demourons/rons ayans le becq fort
Fort suis attaint/tainct de vices peruers
Vers vous rescrips/criz et regretz diuers
Verdz sont les champs/chantz en moy sont arrest
Rest que le pas passer les yeulx couuers
Vers de virlais laisse adieu molinet.

Qui veult scauoir cõe mõ chãt sarroye
Double le quart & la fin de ma roye.

Plate 2.4 Rebus-rhymes for 'Domine mi reverende'; 1531 *Faictz*, fo. V6[v] (Boulogne-sur-mer, Bibliothèque Municipale)

visual techniques. In the *Faictz*, unusual layouts are rarer, and less disruptive of conventional reading processes, than in manuscripts. This presentation reflects a view of rhetoric which accords less importance to the central feature of Molinet's literary self-consciousness, his manipulation of form.

The 1531 *Faictz* thus represents a transitional stage of visual presentation, between the manuscript anthologies permitting close text–image integration and flexible use of space, and the visual uniformity of a more thoroughly typographical layout. The solutions to puzzle-poems typify print's tendency to present books as fixed and complete, rather than permanently unresolved as in the case of manuscripts. Whilst manuscript readers interpreted texts actively, annotating and punctuating them, printers increasingly interpreted texts for readers, as glosses and punctuation became standardized; readers could only receive the text passively.[28] Hence the puzzle-poems' instructions, now redundant as directions to readers, serve as clues to the pieces' composition. The editor pre-empts readers' interpretative activity: we may admire the ingenuity of a poem explained, but need not solve it ourselves. Puzzle- and rebus-poems are presented as finished artefacts for passive consumption, not as challenges to readers as in the major Molinet manuscripts.

The second edition of the *Faictz* continues this trend.[29] Its smaller format, more characteristic of Longis's poetry books, precludes a two-column layout for verse, thereby attenuating the variety of the reader's visual experience. The grammar of legibility is again not especially elaborate: bastard type is used throughout, though, following the first edition, larger type is used for many titles from *Le Temple de Mars*. 'Domine mi reverende' (fo. GG8^{v}) has been completely normalized: the pictorial rhymes, impossible to fit onto the smaller page, have vanished, and each verse now has a full complement of syllables. These alterations further reduce the page's visual expressiveness, and hence reinforce the predominance of moral and historical concerns over formal sophistication.

The 1540 edition retains the changes made in 1537, and introduces roman type and arabic foliation.[30] Its layout continues the trend towards less visual articulation of the text. Some effort is

[28] See Saenger and Heinlen, 'Incunable Description', 253–4.
[29] Bodleian, Douce M 551.
[30] BNF, Rés. Ye 1340.

made to distinguish different textual functions: titles are usually marked with type ornaments different from those used in the corresponding text. In general, however, the visual impression becomes increasingly uniform: a single typeface is used throughout except on the title-page, while *explicits* are rarer. White space is in shorter supply than in previous editions: the *Recollection des merveilleuses advenues*, for instance, lacks any gaps between its stanzas (fos. D7^{r}–G2^{v}), though this tendency is counterbalanced by the lower black-to-white ratio of roman type. Some unusual layouts remain virtually unchanged from 1531,[31] but the broad movement towards plainer text continues.

Hence the *Faictz*, through its selection and ordering of material and its preface, presents Molinet as primarily a didactic poet. Many pieces are visually normalized in comparison with manuscript versions; those unusual layouts which survive are normalized to a degree. These tendencies cannot be wholly explained in terms of economic constraints, precedent, or house style. Clearly, the self-conscious aspects of Molinet's poetry are attenuated; intricate versification and reflexive correspondence pieces both have their place in the anthology, but are very much of secondary importance. Printing is at least partially responsible for the differences between the *Faictz* and the manuscript anthologies: expressive layouts are more difficult to reproduce in print (though this is nevertheless possible), while publishers must address a wide public, not necessarily ignorant but by no means as familiar with *rhétoriqueur* poetics as the more restricted audiences for which manuscripts catered.[32] A new aesthetic is established, which leaves relatively little room for technical or visual experimentation.

The Valenciennes editions

Besides the *Faictz*, published relatively late and in Paris, numerous editions of single Molinet poems were printed in various locations.[33] Several of these pre-date Molinet's death, though his

[31] e.g. *Magistro Ludovico Compere, Ung present* (fo. H3^{r-v}).

[32] Schutz, *Vernacular Books*, 58, identifies six owners of the *Faictz* from notarial inventories in 16th-century Paris. One is Francis I; the others are civil servants and legal officials of varying ranks, and a lawyer's wife.

[33] For details, see Dupire, *Étude*, 98–102, and n. 1 above.

authorship is not always credited. The texts published often exhibit highly sophisticated rhetoric, but are generally of didactic, occasional, or historical interest. This is true of most of the poems printed more than once, such as the *Temple de Mars*, whose nine editions I examine in the next section; the *Testament de la Guerre* and versions of the *Complainte de Grece*, each printed three times; and the *Tres desiree et prouffitable naissance*, printed twice. Up to seven editions appeared in the Burgundian Netherlands: three were printed in Valenciennes, three in Antwerp, and the first edition of the *Temple de Mars* was perhaps produced in the southern Netherlands.[34] These Netherlandish editions exhibit two significant differences from those published elsewhere, mainly in Paris or Lyons. First, the texts they transmit tend to be more clearly pro-Burgundian. Four pieces printed in single-text editions either focus heavily on recent Burgundian history, or manifest Molinet's political loyalties: these are the *Recollection des merveilleuses advenues*, *Complainte de la Terre Sainte* (a reworking, by an anonymous author, of the *Complainte de Grece*), *Robe de l'archiduc*, and *Resource du petit peuple*.[35] All these texts were printed only in the Burgundian Netherlands. By contrast, the texts most frequently printed in France lack any specific political stance: the *Testament* is a rather general satire on war and its effects, while the *Temple* advocates peace without supporting any particular party.[36] Secondly, Netherlandish editions acknowledge Molinet's authorship more often. This tendency is especially striking in the three editions published in Valenciennes: the *Resource du petit peuple*, probably dating from between 1493 and 1500, and two publications from the period 1500–3, the *Robe de l'archiduc* and the first edition of the *Tres desiree et prouffitable naissance*.[37] The *Robe* is explicitly attributed, while Jean de Liège's edition of the *Naissance* bears a pictorial attribution, a woodcut of a windmill with the legend 'molinet' (fo. [1]6^{v}). The surviving copy of the *Resource* is unattributed, but apparently lacks its first leaf;[38] the edition in its original state possibly bore an attribution. Only four other editions are attributed: of these, one (the *Recollection*)

[34] On the *Temple*'s first edition, see G. Colin, 'L'Imprimeur du Flavius Josèphe'.

[35] On these pieces, see respectively C. Thiry, 'Le Vieux Renard et le jeune loup: L'Évolution interne de la *Recollection des merveilleuses advenues*'; Dupire, *Étude*, 123–4; Devaux, *Jean Molinet*, 525, 189–200.

[36] See Devaux, *Jean Molinet*, 480–6.

[37] For dating, see Servant, 'Jehan de Liège', 9.

[38] Colin, 'Jean de Liège', 530.

was printed in Antwerp, and another is the second edition of the *Naissance*, which probably copies the first.

These tendencies are easily explicable. Molinet had no influence over the publication of most editions printed during his lifetime; his influence, which would make explicit attribution more likely, was especially weak outside Burgundian areas. The selection and attribution of printed texts also probably reflect Molinet's reputation as a Burgundian patriot: while a ready market for pro-Burgundian pieces existed in Netherlandish provinces, this was obviously not the case in Paris, where Molinet's very name was perhaps unlikely to help sales. The Valenciennes editions exemplify both these effects. Their frequent attribution suggests that if Molinet indeed collaborated in their publication, as has been widely supposed, this was at least partly for self-promotional reasons. Moreover, all three published texts are of local interest. The *Resource* comments on the war, unrest, and famine besetting the Burgundian Netherlands in the aftermath of Nancy; the *Robe* (*FD*, i. 258–64) recounts events in the same region between Mary of Burgundy's death and the Peace of Senlis in 1493, denoting the Burgundian lands via the metaphor of a fragmented robe; the *Naissance* celebrates the birth of a new generation of the Austro-Burgundian dynasty.

Yet local relevance and authorial self-promotion do not explain why Jean de Liège published these particular texts, and not others. If Molinet indeed co-operated in publication, he could presumably choose which poems to have printed. It is unlikely that he should select pieces of purely local interest, without considering their themes and style, and perhaps how these might be actualized on the page. Analysis of the Valenciennes editions with respect to these issues will suggest how Molinet may have wished his work to be received, at least by a general public.

Probably the first text printed in Valenciennes is the *Resource*, whose political discourse is embellished and nuanced by an intricate *prosimetrum* form and a subtle use of personifications.[39] The presentation of the edition, however, is relatively plain.[40] Decorative initials and type ornaments are absent, and a single typeface used. Layout was probably limited both by early printers'

[39] On the *Resource*'s rhetoric, see A. L. Gordon, '*La Ressource du petit peuple* (1481): Essai de pleine rhétorique'; Armstrong, 'Practice'.

[40] BR, LP 4281 A (facsimile).

tendencies to leave decoration to readers, and by the material available. Jean de Liège doubtless lacked the resources to exploit the visual potential of the concealed alexandrines, which are highlighted only by sporadic *virgulae*: 'Lenfant moult hault crioit par destresse de faim/ La mere se taisoit par traueil inhumain/ lenfant q̃roit sa vie ou sain de sa nourice la mere q̃roit mort ⁊ derrenier supplice' (fo. 3^{r}; *FD*, i. 139, ll. 69–72). The alexandrines have not been overlooked: the *virgulae* signal their presence, but suggest that more striking methods of revealing them, such as different type, are unavailable. Hence the layout stresses the rhyming prose less than do certain manuscripts. The conjunction of technical limitations and the context of publication thus foregrounds the *Resource*'s political content above its formal elaboration, though the latter is hardly likely to escape the reader.

An altogether more unusual layout characterizes the edition of the *Robe*.[41] The poem's stanzas are laid out in a fashion far from universal in early French printing: they are centred on the page, well spaced, and never split over two pages. Consequently, the *Robe*'s versified character is stressed: stanzas are highly discrete, and white space much more copious than for a prose text. Generous margins and spacing produce an aerated layout, counterbalancing the textura typeface's high black-to-white ratio. The centring of stanzas, especially rare in an era when text was normally left-justified, contributes towards visual balance: a left-justified layout would have left large expanses of white space to the right of the short, hexasyllabic lines. All these features suggest the printer's sensitivity to both the *Robe*'s prosody and the reader's visual experience. The edition's production is economical: it lacks a separate title-page, probably so that it may be printed on a single sheet of paper (four quarto leaves), and is decorated only by small *lettres de tourneure* which begin each stanza. Yet despite this economy, Jean de Liège balances the *Robe*'s local appeal with a sensitive presentation. Individually, the unusual features of layout each highlight the *Robe*'s use of verse; together, they visually overdetermine it.

Unlike the other Valenciennes editions, that of the *Naissance* appeared relatively soon after the text's composition.[42] Perhaps

[41] Chantilly, Musée Condé, IV, E. 89 (2).

[42] BNF, Rés. Ye 1077. The poem mentions Charles's christening, on 7 March 1500 (v. 140).

because of its topical relevance, and presumed appeal to both a local market and court circles,[43] there is considerable investment in its publication: it has a separate title-page and three woodcuts. As in the *Robe*, verse is presented sensitively: stanzas are well spaced and never split over two pages, and there is frequent metrical punctuation at internal rhymes (Plate 2.5). More significant, however, are the woodcuts, which require detailed examination.

The title-page bears a cut of a swan, with the legend 'vallenchẽnes' (Plate 2.6). This image may be a printer's mark, with no specific link to the *Naissance*.[44] Nevertheless, it at least clearly situates the edition in a milieu befitting the poem's Burgundian patriotism. Moreover, it reflects and draws attention to Molinet's use of language in the *Naissance*. The swan is a visual metaphor for Valenciennes, based on an etymological pun which Jean Lemaire later exploited in his *Épitaphe de Molinet*:

> —Comment a nom ce lieu qui telz les a nourry?
> —Val en cignes val doulx val insigne et floury.[45]

The legend's orthography, unlike that of Lemaire's poem, does not immediately suggest swans; however, the juxtaposition of 'vallenchẽnes' with the swan image provokes readers into making the connection and recognizing the visual pun. Hence readers become sensitized to the possibility of similar wordplay within the poem. They are thus more likely to appreciate the punning with which Molinet praises Charles's birthplace, employing *gand* with the simultaneous meanings of 'glove' and 'Ghent':

> Gant, tres fort Gant, Gant tres doulx et humain,
> En qui la main de Dieu se loge et boute. (vv. 65–6)

The pictorial attribution on the final page (Plate 2.7) provides another pun, this time based on Molinet's characteristic punning signature. Its legend 'molinet' refers to both the windmill and Molinet's name, while the image of the windmill both attests this

[43] Brown, *Poets, Patrons, and Printers*, 129, suggests that Molinet perhaps chose to distribute the poem in printed form.

[44] Only the place of publication is shown: nothing links the cut specifically to Molinet or the *Naissance*, and it could conceivably be reused in other publications by the same printer, though no recurrences are known.

[45] *Œ*, iv. 320. Further references to this edition will be incorporated into the text. On Lemaire's epitaph, see Jodogne, *Jean Lemaire*, 285–9.

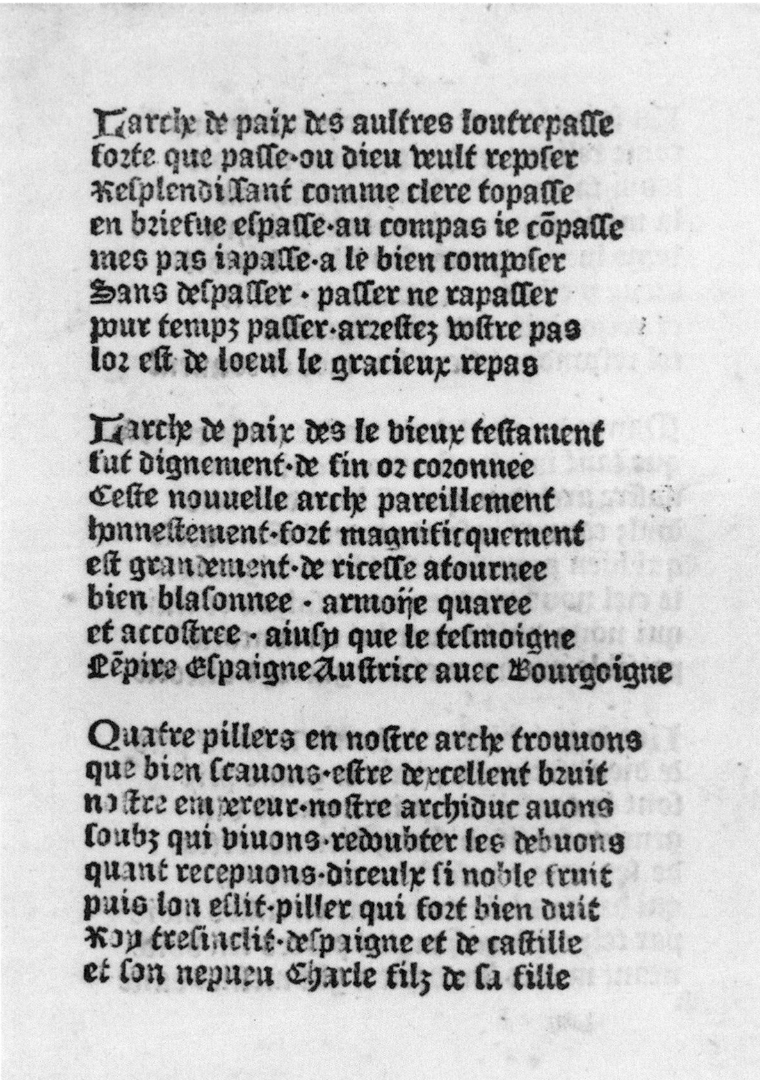

Larche de paix des aultres soutrepasse
forte que passe·ou dieu veult reposer
Resplendissant comme clere topasse
en briefue espasse·au compas ie cōpasse
mes pas iapasse·a lē bien composer
Sans despasser · passer ne rapasser
pour tempz passer·arrestez vostre pas
lor est de loeul le gracieux repas

Larche de paix des le vieux testament
fut dignement·de fin or coronnee
Ceste nouuelle arche pareillement
honnestement·fort magnificquement
est grandement·de ricesse atournee
bien blasonnee · armoirie quaree
et accostree · ainsy que le tesmoigne
Lēpire Espaigne Austrice auec Bourgoigne

Quatre pillers en nostre arche trouuons
que bien scauons·estre dexcellent bruit
nostre empereur·nostre archiduc auons
soubz qui viuons·redoubter les debuons
quant receuons·diceulx si noble fruit
puis lon eslit·piller qui fort bien duit
Roy tresinclit·despaigne et de castille
et son nepueu Charle filz de sa fille

PLATE 2.5 Metrical punctuation in the Valenciennes *Naissance*, fo. [1]2^r (Cliché Bibliothèque Nationale de France, Paris)

PLATE 2.6 Title-page of the Valenciennes *Naissance* (Cliché Bibliothèque Nationale de France, Paris)

first possible meaning and continues the pun in the closing lines of the text, where Molinet wishes that

> sans amere. orge ait en son van net
> grain et bon vent. vostre humble molinet.
> (fo. [i]5^{v}; *FD*, i. 358, vv. 183–4)

PLATE 2.7 Pictorial attribution in the Valenciennes *Naissance*, fo. [1]6^{v} (Cliché Bibliothèque Nationale de France, Paris)

By depicting a windmill, the woodcut ensures this pun is not read as a dead metaphor, but recognized as a conjunction of proper name and common noun; we cannot read 'molinet' in the text as the poet's name alone. The woodcut thus enables Molinet's name to be absorbed into the text as signifying system: 'molinet' can be integrated to the poem's closing agricultural image, and thus made to signify, rather than simply to denote a referent in the person of Molinet. Consequently, the cut helps convey the density of Molinet's language, where no element is unmotivated.

A later edition of the *Naissance*, printed in Lyons, reveals that the Valenciennes edition's windmill image succeeded in identifying Molinet as author.[46] This second edition's title is strikingly similar to that of the first, suggesting that Jean de Liège's edition was copied. However, the Lyons edition's title also mentions Molinet by name: '¢ Cy cõmence la tresdesiree et prouffitable naissance du tresillustre enfant Charles dautriche filz de trespuissant prince mõseigneur larcheduc tresredoubte prince/ Laquelle natiuite a este cõposee par vng fatiste appelle ¢ Molinet.'[47] The later printer has clearly recognized that the windmill constitutes an attribution to Molinet.[48]

The third woodcut in the Valenciennes *Naissance* visually represents the poem's governing metaphor, the 'arche de paix' (fo. [1]1^v; Plate 2.8). Accompanying the ark are the arms of the Empire, Austro-Burgundy, and Spain, denoting the lands

[46] n.d.o.p. [Lyons, Guillaume Balsarin, *c.*1503] (BNF, Rés. Ye 221); for dating and location, see H. Baudrier, *Bibliographie lyonnaise* (Lyons, 1895–1921), xii. 51.

[47] fo. d1^r; for the Valenciennes edition's title, see Plate 2.6.

[48] Brown, *Poets, Patrons, and Printers*, 131–3, notes the Lyons edition's explicit attribution, and the presence of a woodcut depicting the author at his desk. She considers that these features increase Molinet's authorial prominence. However, this edition was apparently not printed separately. It consists of a single gathering signed d, which according to Anne Rouzet forms the second part of a volume entitled *Les Croniques des roys, ductz et contes de Bourgoigne* (BR, Inc. A 409 LP; see P. Cockshaw, C. Lemaire, and A. Rouzet, *Charles le Téméraire: Exposition organisée à l'occasion du cinquième centenaire de sa mort* (Brussels, 1977), 149–50). Indeed, the *Croniques* is signed a–c, and uses the same characters as Molinet's piece, which is mentioned on its title-page, though without any mention of Molinet's name: '¢ Item est compris auec lesdites croniques la tresdesiree τ prouffitable naissance du tresillustre enfant Charles dautriche filz de trespuissant prince mõseigneur larcheduc.' Hence, while the *Naissance*'s separate title-page may have enabled it to be sold separately for reasons of commercial flexibility, it originally formed part of a more general project. Molinet's prominence within the *Naissance* is thus not borne out in the whole volume.

PLATE 2.8 The 'arche de paix' in the Valenciennes *Naissance*, fo. [1]1v (Cliché Bibliothèque Nationale de France, Paris)

and dynasties united in the person of Charles, as the poem indicates:

> Ceste nouvelle arche . . .
> Est grandement de ricesse atournee,
> Bien blasonnee, armoyee, quaree
> Et accostrée, ainsy que le tesmoigne
> L'empire, Espaigne, Austrice avec Bourgoigne. (vv. 11–16)

This link between the ark and Charles is reinforced in the poem by anagrammatic wordplay (*l'arche/ Charles*), and in the woodcut by the arches which surmount the ark, recalling the appearance of the imperial crown.[49] The allusion to a *couronne à arches* intensifies the punning, and hints at Charles's imperial inheritance.

However, elsewhere the ark represents not Charles, but his mother Joanna the Mad:

> nostre archiducesse,
> Haulte princesse, est de paix la droite arche. (vv. 76–7)

This alternative reading is supported by medieval theology, which saw the Ark of the Covenant as one of the Emblems of the Virgin; the ark is thus appropriate for a mother rather than her child.[50] Hence the metaphor suggests a flattering parallel between Charles and Christ, which the poem reinforces by comparing Charles's birth with the Nativity, and noting that both are associated with peace:

> Le filz de Dieu qui le monde vainqui
> En paix nasqui dessoubz Octoviien;
> L'an mil cinq cens, ung mains, fevrier, rendi,
> Par ung lundi que bissexte escheÿ,
> Ce filz icy; soubz Maximiliën,
> Nous sommes bien. (vv. 105–10)

In other ways, too, the ark metaphor oscillates between Joanna and Charles. The 'Quatre pillers' supporting the ark (vv. 17–24) are identified as Maximilian, Philip, Ferdinand of Spain, and the infant Charles, implying Charles cannot also be the ark itself, yet the numerous verbal and visual puns imply the contrary. Moreover, the explicit interpretation of the ark as Joanna does not accord with the uniting of the four dynasties: unlike Charles, Joanna is linked to Burgundy, Austria, and the Empire only by marriage, not by blood.

The woodcut's obvious contribution to this ambivalence is reinforced by its position between title-page and text, which provokes readers into attempting to interpret it before reading the text. As the title suggests the cut bears some relation to Charles,

[49] See R. W. Scheller, 'Imperial Themes in Art and Literature of the Early French Renaissance: The Period of Charles VIII', 9–10.

[50] Réau, *Iconographie*, II. i. 207.

readers are likely to interpret it as a metaphor for him, a reading supported by the coats of arms. Textual details make us revise this interpretation, which, however, remains partially valid since so many verbal and visual allusions favour it. Consequently, the cut highlights the shifting nature of the ark's figurative meanings. Jean de Liège's presentation of the *Naissance* thus accentuates two of Molinet's self-conscious techniques. Metrical punctuation and woodcuts draw attention to Molinet's manipulation of language through complex versification, puns, and elusive metaphors. Simultaneously, the pictorial attribution gives Molinet a prominence which reflects the apparently authorial first-person voice in his reference to 'vostre humble molinet'.

All three Valenciennes editions, to varying degrees, display three major traits. First, their generally clear attributions and local appeal suggest Molinet may have co-operated in publication; second, Molinet's Burgundian patriotism is underlined; thirdly, formal features are often highlighted. The editions therefore achieve a subtle balance between the political and the formalistic. However, they are aimed at a relatively mass readership, not perhaps fully conversant with prosodic intricacies. Their priorities are thus not necessarily those followed when presenting a poem to the members of one's poetic community, patrons or fellow-poets.

Le Temple de Mars

Molinet's most frequently printed piece was the *Temple de Mars*, an anti-war poem probably composed in late 1475. The eponymous building plays a relatively brief role in the poem, stressing war's disruption of proper order by presenting it in terms of religious ritual:[51]

> Le chant de ce temple est alarme,
> Les cloches sont grosses bombardes. (vv. 41–2)

Versification is characteristically sophisticated: the *Temple* is in alternating decasyllabic and octosyllabic *huitains*, of which the

[51] Cornilliat, *'Or ne mens'*, 660–75, examines the way in which the *Temple* both employs such rhetoric and suggests its limitations.

former use *rimes batelées*. Certain stanzas employ further prosodic refinements, such as *rimes couronnées* or *annominatio*:

> Guerre a fait maint chastellet let
> Et mainte bonne ville ville; (vv. 169–70)
>
> Par guerre sont marchans meschans,
> Gentils chetifz, pecheurs pieurs. (vv. 185–6)

Cynthia Brown has effectively analysed some key tendencies in the *Temple*'s publication, such as the lack of attribution in single-text editions, and the use of woodcuts and other paratextual features which appear to glorify war, thus distorting Molinet's anti-war message.[52] The following discussion complements Brown's approach, and focuses primarily upon the ways in which editions present the text's imagery and versification.

The first *Temple* edition appeared during the infancy of French-language printing.[53] Its plain presentation is typical of this period: it lacks a separate title-page and decorative initials. Decoration and punctuation have been carried out by a reader of the only surviving copy, who has painted a red initial over the guide letter which begins the text; added a red paraph and *bâton* to the title; stroked the first letter of each stanza in yellow ink; and frequently punctuated the text with *virgulae*. However, these annotations reveal no readerly interest in the *Temple*'s formal characteristics: *virgulae* have not been inserted after internal rhymes or between the *rimes couronnées*, as sometimes occurs in manuscripts.[54]

All but one of the subsequent *Temple* editions appeared in Paris. They are more elaborate than the first, with separate title-pages, and woodcuts in most cases. However, their printers make no more use of systematic metrical punctuation than the reader just mentioned.[55] Moreover, the woodcuts are usually stock images, whose relevance to the text is tangential at best. The second edition, for instance, bears a cut on the title-page verso, depicting

[52] Brown, *Poets, Patrons, and Printers*, 66–71.

[53] n.d.o.p. [*c.*1476] (Musée Condé, IV, G. 15). For dating, see *Recueil*, ed. Picot and Stein, 31–3.

[54] In the Cambridge manuscript, frequent *virgulae* at the hemistich highlight the *rimes batelées* in the *Temple* (pp. 78–88); in BR, ms. II, 2545, some of the *rimes couronnées* are marked by *punctūs* or *virgulae* between the repeated syllables (fo. 278[r]).

[55] This is typical of single-text editions printed outside Valenciennes, though *rimes batelées* are highlighted in *La Complainte de Constantinoble*, a version of the *Complainte de Grece*, n.d.o.p. (BL, 11475 d. 31 [now missing]). Here, verses containing *rimes batelées* are split at the caesura and laid out over two lines: Dupire, *Étude*, 123.

a knight kneeling before a king, being crowned by a lady as other knights and ladies watch.[56] While the king is labelled 'Mars. Dieu des batailles', the scene bears no relation to the text: even if it were cut especially for this edition, its designer was either ignorant of the text or deliberately produced a generic image which, with suitably altered legends, could be re-employed in subsequent publications. The cut does suggest a temple, as it depicts a ceremony inside a large building; however, this temple is most unlike that in the text, which is a place of chaos and misery, not of court ceremony. The image's design was probably based on the poem's title alone, which offers the motifs of Mars and the temple without indicating the ways in which Molinet elaborates them.

Later Paris editions similarly employ cuts which have little in common with Molinet's poem. Images include a tournament;[57] a king with three knights;[58] an emperor receiving a document;[59] and a castle keep with a woman at the ramparts.[60] The use of cuts representing rulers and buildings implies that printers have sought to depict Mars or the temple, but without considering their role in the text.[61] Lack of printerly attention and investment is also reflected by cramped layouts: there is little white space, and two editions have no gaps between stanzas.[62] These features are characteristic of ephemera. The reuse of woodcuts, in particular, typifies the illustration of such texts as reports of royal entries or battles; the tournament, for instance, was a common motif in the illustration of information sheets.[63] Indeed, Cynthia Brown has

[56] Paris, Le Petit Laurens, n.d. [*c.*1491] (New York, Pierpont Morgan Library, PML 75124); reproduced in Brown, *Poets, Patrons, and Printers*, 68.

[57] Paris, Jean Trepperel, n.d. [*c.*1497–8] (BNF, Rés. Ye 1127), fo. A1^{v}; reproduced in Brown, *Poets, Patrons, and Printers*, 69.

[58] Paris, no pub., n.d. [Alain Lotrian, 1526–30] (BNF, Rés. Ye 1282), title-page; for dating and location, see Moreau, *Inventaire*, iii. 1069 *bis*.

[59] Paris, no pub., n.d. [Alain Lotrian, *c.*1531] (Paris, Bibl. de l'École Nationale Supérieure des Beaux-Arts, Rés. Masson 469), title-page. On this edition, see Moreau, *Inventaire*, iv. 245.

[60] Paris, no pub., n.d. [Alain Lotrian, *c.*1520] (BNF, Rés. Y^{2} 2579), title-page.

[61] Several other Molinet editions also employ stock cuts with only tangential relevance to the text. Examples include the title-pages of editions of the *Testament de la Guerre*, which bear woodcuts depicting soldiers: Lignan, no pub., n.d. (BNF, Rés. Ye 1376); n.d.o.p. [Paris, *c.*1520] (Musée Condé, IV, D. 92); Paris, widow Nicolas Buffet, 1559 (Aix-en-Provence, Bibl. Méjanes, Rés. S 44 (4)).

[62] n.d.o.p. [Paris, Michel Le Noir, 1501–5] (BNF, Rés. Ye 220); n.d.o.p. [Paris, Jean Trepperel, 1506–9] (BNF, Rés. Ye 273).

[63] J.-P. Seguin, *L'Information en France, de Louis XII à Henri II* (Geneva, 1961), 10–17.

suggested that Parisian editions of the *Temple* were often printed to comment upon contemporary wars involving France: their illustrations, which present war in more positive terms than does the poem, perhaps reflect printers' propagandist attitudes regarding French military expeditions.[64] It is clearly the printers' perception of the *Temple* as ephemera, rather than limited resources, which determines the plainness of presentation: well-established and prolific publishers such as Jean Trepperel, Michel Le Noir, and Alain Lotrian were hardly short of material. Printers were more interested in the *Temple*'s perceived relationship to actuality than in its formal qualities.

A more sensitive presentation of the *Temple* characterizes an edition published in Lyons in 1502.[65] This is the only edition to exploit the potential for illustration provided by Molinet's description of Mars:

> Mars triumphoit en son noble carroy,
> Ainsy que ung roy armé de pied en cappe;
> Trembler faisoit murs, chasteaux et beffroy
> Par son effroy et tenoit, sans arroy,
> En tout desroy, flaÿel dont nul n'escappe. (vv. 65–9)

On each side of the first leaf, a woodcut depicts a figure in a chariot, holding a sword and severed head (Plate 2.9). The chariot motif clearly constitutes a close textual parallel, while the zodiacal signs Aries and Scorpio, on the chariot's wheels, were recognized iconographical attributes of Mars and thus enable knowledgeable viewers to confirm the figure's identity.[66] Indeed, the cut was probably designed especially for the *Temple*. No previous uses are attested; though the motif of Mars's chariot appears in Petrarch's *Africa* and elsewhere,[67] this cut is no closer to *Africa*'s text than to the *Temple*'s. There is no reason to assume it has been designed for any other edition than this one.

However, the cut is not a wholly faithful illustration. Mars

[64] Brown, *Poets, Patrons, and Printers*, 66–71.

[65] Lyons, no pub. [Jean de Vingle], 18 December 1502 (BNF, Rothschild 2580); for the printer, see *Recueil*, ed. Picot and Stein, 39–41.

[66] Tervarent, *Attributs*, 70, 395–6; Aries and Scorpio are Mars's ascendant and descendent houses.

[67] Mars has a helmet, a whip or other weapon, and is accompanied by a wolf; two horses draw his chariot. This representation is also standard in visual images: see J. Seznec, *La Survivance des dieux antiques: Essai sur le rôle de la tradition mythologique dans l'humanisme et dans l'art de la Renaissance* (London, 1940), 165–8, and pl. 74–9.

PLATE 2.9 Title-page of the Lyons *Temple de Mars* (Cliché Bibliothèque Nationale de France, Paris)

carries a sword, not a 'flaÿel'; he is not 'armé de pied en cappe', but has a bare torso. Similarly, the animals surrounding Mars have few textual parallels. The chimera perhaps represents the figure of Guerre, described as 'une laide chimere' in the poem (v. 98), but it lacks various traits which Molinet ascribes to Guerre: it has neither a 'gueulle draconicque' nor a 'queue d'escorpion' (vv. 116, 121). The eagle recalls an episode near the beginning of the *Temple*, in which the first-person narrator offers various animals to Jupiter, including 'Ung aigle d'estrange plumage' (v. 29). Yet other motifs in the cut are absent from the text. Iconographic precedents explain the relevance of some elements, such as Aries and Scorpio; the eagle, a traditional symbol of power; and the crayfish, which could denote bad fortune.[68] But others cannot be interpreted with certainty despite having accepted iconographical meanings, because these do not explicitly appear in the *Temple*. The winged horse may represent fleeting fame,[69] reinforcing the text's suggestion that antique heroes, 'Paris, Achillés, Troÿlus' and the rest (v. 273), were rewarded for their warlike endeavours only by death. Similarly, the bull and scales probably represent Taurus and Libra, the ascendant and descendent houses of Venus, but the relevance of this to Molinet's poem is uncertain. The implication may be that war and peace alternate with a regularity comparable to planetary movements; however, this interpretation cannot be confirmed by the text.

The cut, therefore, cannot easily be mapped onto the poem. While Mars is easily identifiable, it is often difficult to ascribe univocal meanings to other pictorial elements. Consequently, readers must enquire into the detail and implications of the *Temple*'s metaphors. Faced with a woodcut obviously depicting Mars, we are tempted to interpret its diverse elements by relating them to the text. Some motifs, however, have no explicit textual motivation, and provoke readers into constructing their own interpretations, attempting to create meaningful connections between these elements and Mars on the basis of the text's presentation of war. Hence the woodcut stimulates us to read the *Temple* closely. We must assess its moral import, judging which aspects of Molinet's critique of war might motivate certain pictorial motifs. Indeed, the cut's very presence suggests that Mars is important to the poem: readers cannot simply read the *Temple* as a moral (much less a propagandist) piece, without attending to Molinet's use of *poetrie*. Furthermore, text–image anomalies highlight some of the implications of Molinet's

[68] For the latter two motifs, see Tervarent, *Attributs*, 23–4, 187–8.
[69] Ibid. 119–20.

description. Mars in the woodcut, for instance, is not fully armoured 'Ainsy que ung roy' as in the text, a discrepancy which raises the question of why Mars should be compared to a king at all. The answer, suggested by the *Temple*'s pleas for peace, is that rulers often wage war without considering the consequences.[70] The woodcut thus provokes both an appreciation of the *Temple*'s lessons and a recognition of the allegory's importance in the work.

This edition, however, is exceptional; Parisian editions of the *Temple* place little emphasis on the poem's self-conscious features.[71] The absence of attribution, and of elements which foreground Molinet's use of versification and allegory, reflects the way in which printers view the text as relating primarily to actuality.[72] Other single-text editions of Molinet's poetry evince similar priorities: the title-pages of two *Testament* editions, for example, refer to 'le testamẽt de la guerre Qui regne a p̃sent sur la terre'.[73] Moreover, indications of use reveal that readers sometimes shared printers' attitudes. The surviving copy of the first *Temple* edition is a case in point: on its blank final page, a contemporary hand has added historical notes in Latin for the years 1487–96, with entries such as 'Exultatio Rusticorũ' for 1489. Such annotations clearly suggest that the *Temple* was treated as a comment on current and recent events.

Conclusions

Editions of Molinet's poetry fall into three categories. The Valenciennes editions balance politics and formal self-consciousness, giving Molinet a presence in the books and accentuating his Burgundian credentials. Other single-text editions, exemplified by those of the *Temple*, tend to attenuate the poetry's self-conscious aspects: presentation is often characteristic of ephemera, while

[70] Cf. esp. vv. 73–80.

[71] Brown, *Poets, Patrons, and Printers*, 71, suggests that the contrast between the Lyons edition and those printed in Paris reflects the more intense competition among Parisian printers.

[72] The *Temple* is, however, attributed in a small anthology of *rhétoriqueur* poetry, the *Traictez singuliers*. This was published three times, but only the first two editions contain the *Temple*: Paris, Antoine Couteau for Galliot Du Pré, February 1525/6 (BNF, Rothschild 487); Paris, for Galliot Du Pré, 'February 1525' (Arsenal, Rés. 8° BL 8749). See Cretin, *Œuvres*, p. xcv.

[73] BNF, Rés. Ye 1376; see also Musée Condé, IV, D. 92 (for details of editions, see n. 61 above).

attributions are rare. The *Faictz* presents Molinet's work as primarily didactic, historical, and political, though his Burgundian sympathies are moderated. Formal virtuosity is acknowledged, but not granted its full importance in Molinet's practice. In general, then, printers are less apt than scribes to reflect poems' formal features visually. Even the *Faictz*, which contains several poems with expressive layouts, reduces the extent to which these pieces challenge readers to reconstruct their meaning actively.

The difference between manuscripts and editions in this respect can fruitfully be understood in terms of the distinction drawn by Roland Barthes between two types of text, the *texte scriptible* and the *texte lisible*. Whereas the reader of the former is 'non plus un consommateur, mais un producteur du texte', the *texte lisible* grants readers only 'la pauvre liberté de recevoir ou de rejeter le texte'.[74] If Barthes's terms are regarded not as rigid categories, but as poles in relation to both of which any text may be situated in the course of its reception, it becomes clear that editions of Molinet's poetry generally present it as less *scriptible* and more *lisible* than manuscripts suggest.

Furthermore, Molinet's works are put to a narrower range of uses in print than in manuscripts. This is perhaps inevitable, since printed books could not address individual readers' preferences in the same way as manuscripts. Significantly, printers' selections reveal a reception of Molinet centring on specific areas. The Valenciennes editions focus on local affairs, addressing a pro-Burgundian readership without neglecting Molinet's rhetoric. As the *Temple* editions show, most single-text editions printed outside Valenciennes privilege texts' perceived contemporary relevance. Certain strands of Molinet's output are practically non-existent in print. Of his explicitly bawdy verse, for instance, there is none in the *Faictz*; only the *Complainte d'un gentilhomme* is printed separately,[75] while the rondeau 'Ceste fillette a qui le tetin point' (*FD*, ii. 875) appears in several printed multi-author collections, albeit with no acknowledgement of Molinet's authorship.[76]

Differences between the manuscript and printed transmissions of Molinet's poetry seem related to the modes of literary production and reception which the two media encourage. Manuscripts can be constructed according to idiosyncratic preferences,

[74] R. Barthes, *S/Z* (Paris, 1970), 10.

[75] For editions, see Lachèvre, *Recueils*, 181.

[76] Dupire, *Étude*, 104.

while printing requires a mass market; scribes are freer than printers to represent formal patterns iconically. The media seem to appeal to different publics: printers and their audiences are not generally concerned with the self-conscious features of Molinet's verse, whereas his rhetoric is praised by other *rhétoriqueurs*, who presumably knew his work from manuscripts.[77]

Positing these different audiences has important implications for the Valenciennes editions. They exploit printing's capacity to diffuse texts to a wide (and politically sympathetic) readership, but also stress prosody, language, and metaphor in a manner more characteristic of manuscripts. Consequently, they solicit both potential audiences, maximizing their commercial appeal. They effectively convey various aspects of literary self-consciousness, both in highlighting formal features and in publicizing Molinet's authorship. These editions' clear correspondence to both Molinet's material interests and his poetic priorities provides, if not conclusive proof, then at least a strong case in favour of concluding that Molinet indeed co-operated in their publication.

[77] On poets' praise of Molinet, see Dupire, *Jean Molinet*, 1–4. On the different audiences of manuscript and printed versions of the same text in this period, see also M. Camille, 'Reading the Printed Image: Illuminations and Woodcuts of the *Pèlerinage de la vie humaine* in the Fifteenth Century', 270–4.

3

MANUSCRIPTS OF LEMAIRE'S POETRY

Introduction

Though his career partly overlapped with Molinet's, Jean Lemaire de Belges belonged to a younger generation of poets, to whom printing was more familiar and who were apt to exploit the new medium.[1] Accordingly, printing plays a greater role in transmitting his works than it did for Molinet; indeed, as I discuss in the next chapter, Lemaire apparently participated in publication on several occasions. Yet his work also survives in numerous manuscripts. Several of these exhibit features which, in conjunction, suggest they were probably presentation manuscripts: they are of high material quality (written on parchment, illustrated, or both), and transmit pieces by Lemaire in versions which pre-date their printed redactions, or which were never printed in the sixteenth century.[2] These volumes' relatively high incidence is probably not simply because they are more likely to have survived than cheap paper manuscripts; it also suggests that manuscripts were being produced more rarely in the early sixteenth century, so that books confected for presentation to patrons constituted a higher proportion of the total. The vicissitudes of Lemaire's career also probably explain the high proportion of presentation manuscripts: after the death of his employer Louis of Luxembourg in 1503, he sought another patron, and was eventually engaged by Margaret of Austria, becoming the official Burgundian *indiciaire* after Molinet's death. In 1511–12, he left Margaret's service for Anne of Brittany, wife of Louis XII of France; little is known of Lemaire after the deaths of Anne and Louis in 1514 and 1515. Manuscripts, then, are both quantitatively

[1] Such attitudes are evinced by André de La Vigne (*c.*1470–*c.*1515) and Pierre Gringore (*c.*1475–*c.*1539). For details of these authors' involvement in their works' publication, and of its effects on the presentation of their writing, see Brown, *Poets, Patrons, and Printers*, 1–2, 17–19, 34–8, 79–97, 110–16, 133–50, 166–93.

[2] Such manuscripts include three volumes examined in this chapter: Carpentras, Bibl. Inguimbertine, ms. 412; St Petersburg, State M. E. Saltȳkov-Shchedrin Public Library, MS Fr. F. v. XIV, 8; and ÖNB, Cod. 3441. Other possible presentation manuscripts include

and qualitatively important in the transmission of Lemaire's poetry.

Lemaire's poetry differs from Molinet's not only in its modes of transmission, but also in the forms of literary self-consciousness it displays. While Molinet's verse was overwhelmingly characterized by ostentatious formal virtuosity, Lemaire's displays a wider range of self-conscious features. Ingenious versification is present in some poems, most notably 'Salut a vous dame de hault paraige', an address to the Virgin into which the Latin *Salve Regina* is incorporated as an acrostic, each line of verse containing one syllable of the *Salve*.[3] In a departure from the usual configuration of acrostics, these syllables vary in position, combining to form different visual patterns in each stanza: they may form a vertical line along a stanza's left-hand edge, diagonal lines running in different directions in successive stanzas, and so forth.[4] In other pieces, however, self-consciousness is manifested as the overt display of poems' relationships with the poetic tradition or intertext off which they play. This is especially striking in the *Épître du roi Louis XII à Hector de Troie* (*Œ*, iii. 68–86), written in November 1511 in response to Jean d'Auton's *Epistre faicte et composee aux champs Elisees par le preux Hector de Troye la grand transmise au trescrestien Roy Loys douziesme de ce nom.*[5] D'Auton's poem, probably composed in August 1511, whimsically adopted Hector's voice to praise Louis for his victories in the Italian wars. Lemaire's reply, in Louis's voice, not only provides an account of France's Trojan heritage, but also strongly criticizes Pope Julius II for his alleged bellicosity and indifference

BNF, mss. fr. 1683 and 23988, which both transmit a pre-print version of the *Plainte du Désiré* (both volumes are illustrated, and the latter is on parchment: see J. Lemaire de Belges, *La Plainte du Désiré*, ed. D. Yabsley (Geneva, 1932), 60–1), and BNF, ms. fr. 25295, an illustrated parchment manuscript of political pieces never printed in the 16th century (see Jodogne, *Jean Lemaire*, 386–7). Munn cites a total of twenty-four manuscripts containing Lemaire's poetry or historical writing. To these should be added the Carpentras and St Petersburg volumes; BR, ms. IV, 630/14 (see P. Jodogne, 'Le Fragment manuscrit Bruxelles, B.R., IV 630/14: Une page des *Illustrations de Gaule* de Jean Lemaire de Belges'); and Glasgow, Hunterian Museum, MS 12 (S. 2. 2) (see A. Adams, 'Le Manuscrit huntérien de l'*Epistre du roy* de Jean Lemaire de Belges et le manuscrit B.N. fonds français 1690').

[3] Edited in P. Jodogne, 'Un recueil poétique de J. Lemaire en 1498', 204–7.

[4] For discussions of this layout, see F. Rigolot, *Le Texte de la Renaissance: Des rhétoriqueurs à Montaigne* (Geneva, 1982), 52–5, and Gros, *Poète marial*, 79–85.

[5] No modern edition of d'Auton's piece exists; all quotations are taken from the text in Saltўkov-Shchedrin Library, MS Fr. F. v. XIV, 8, fos. 66^{v}–81^{r}. Both Lemaire's and d'Auton's poems will be included in a forthcoming edition of Lemaire's polemical works from the period 1511–13, edited by A. Armstrong and J. Britnell.

to the faith, and Venice for its impiety and perfidy.[6] Yet the *Épître*'s polemical value is complemented by close echoes of d'Auton's poem, which suggest that Lemaire is responding to and developing d'Auton's comic anachronisms.[7]

Louis's assurance that Hector is not necessarily condemned to damnation because he was a pagan, for example, wittily highlights the incongruity between the mythological conceit and contemporary reality:

> Tu doncques, puis qu'en nulle place et lieu,
> N'as fait offense au saint peuple de Dieu,
> Ne dois iamais douter qu'on te reproche
> Si ta creance à la nostre n'approche. (*Œ*, iii. 70)

Similarly, Lemaire's description of modern warfare, characterized by artillery, tacitly suggests that Hector's brand of military prowess is somewhat behind the times:

> Tu ne veis onc si estrange deluge:
> Car de ton temps, les guerres, et victoires
> On les faisoit en bras fulminatoires
> Tant seulement: Mais nostre artillerie,
> Sans point de faute, est vne diablerie:
> Car posé or, que tu, plus corpulent
> Beaucoup que nous, et de coeur excellent,
> Penses reuiure, et reprendre les armes . . .
> ton grant corps seroit plustot atteint
> Qu'vn plus petit, qui sassouplist, ou feint. (*Œ*, iii. 75)

Most telling, however, is Louis's surprise at Hector's fluent French, which points up the absurdity of the very notion of a vernacular correspondence involving a mythological figure:

> Mais quel' merueille? et qui eust eu pensee
> Que nostre langue ainsi propre, agensee
> Fust ia commune en ta tresnoble court . . . ? (*Œ*, iii. 69)

There is a strong sense of literary competition: Lemaire's poem is not solely propagandist, but takes up the literary challenge implicit in Hector's request to Louis to 'me donner sur ce quelque response' (fo. 80^{v}). Indeed, Lemaire simultaneously works within

[6] See Jodogne, *Jean Lemaire*, 395–403; J. Beard, 'Letters from the Elysian Fields: A Group of Poems for Louis XII', esp. 30–2.

[7] See LeBlanc, *Va lettre va*, 113–16.

and punctures d'Auton's comic conceit, demonstrating his ingenuity by giving a further twist to the scenario established by his fellow-poet.

Still other poems thematize poetic composition. Notable in this respect is the verse prologue to the *Couronne margaritique*, a long *prosimetrum* allegory, which ostensibly commemorates Philibert of Savoy (d. 1504), but is largely devoted to praising his widow and Lemaire's current protector, Margaret of Austria.[8] In the prologue, a first-person voice bewails his obligation to write yet another commemorative piece, to

briser mon doux estude,
Laisser escrits de noble claritude,
Pour la plonger es profondes tenebres
De cas diuers, violents et funebres,
Pleins d'infortune, accumulez de deuil,
Lardez de pleurs, farcis de larmes d'oeil. (*Œ*, iv. 15)

The writer obliquely alludes to his previous funereal works by claiming that he must

en tristeur manier
Mon papier noir, et *derechef* le teindre
De grans regretz. (*Œ*, iv. 16; emphasis added)

Readers are tempted to regard these verses as alluding to Lemaire's previous commemorative pieces, the *Temple d'Honneur et de Vertus*, in memory of Pierre of Bourbon, and the *Plainte du Désiré*, on the death of Louis of Luxembourg.[9] Hence we are encouraged to conflate the voice in the prologue with that of Lemaire the historical author. Even more suggestively, the prologue is addressed to the poet's 'Plume infelice', whose tyranny is criticized:

Dites pourquoy mon engin peu fertile
Vous retirez de son emprise vtile
Pour le tourner en ce present traueil? (*Œ*, iv. 15)

Both poet and pen are autonomous beings. That the former can address the latter suggests both the independence of the writing process, and the writer's awareness of that process; a potent image of writerly self-consciousness.[10]

[8] *Œ*, iv. 1–167. Jodogne, *Jean Lemaire*, 215–54, discusses the *Couronne*.

[9] On these pieces, see Jodogne, *Jean Lemaire*, 170–214.

[10] Perhaps significantly, the prologue is transmitted separately, albeit anonymously, in BNF, ms. fr. 1721 (fo. 104^{r-v}).

While not exhaustive, these examples illustrate the variety of self-conscious elements in Lemaire's poetry. In this chapter, I shall examine four important manuscripts, which each foreground different self-conscious techniques. I begin with the earliest known evidence of Lemaire's literary activity, a manuscript of 1498, whose anthologization and illustration highlight Lemaire's use of language and versification, and the poetic context into which he inserts himself. Subsequently, I study a manuscript containing the *Épître du roi*, which draws attention to both Lemaire's play upon d'Auton's piece and his status as court poet; a manuscript of the *Concorde des deux langages*, whose illustrations help convey the sophistication of the text's allegory; and a manuscript of the *Couronne margaritique*, which accentuates both the prominence of Lemaire as historical author and the text's characteristic reflexive elements.

Delineating a literary stance: BNF, n. a. fr. 4061

BNF, n. a. fr. 4061 is a highly personal volume, commissioned by Lemaire himself; it includes two Lemaire poems, and various pieces and extracts by other writers.[11] A note identifies the book's owner and date of confection:

> Ce petit liuret sommaire.
> de la main maistre Regnault.
> Appartient a Iehan le Maire.
> Ne du pays de haynault.
> De riches motz et grant sens.
> chũn voit. quil nest pas vuid
> Escript lan mil quatre cens.
> Quatre vingtz. et dix huit/. (fo. 2^{v})

Three significant aspects of Lemaire's writerly practice are exemplified here. First, the explicit references to the volume's owner and scribe evince Lemaire's interest in the materiality of texts, an interest later manifested in the illustration of presentation manuscripts, and in Lemaire's involvement in printing. Indeed, this manuscript is well produced, surprisingly so in view of Lemaire's relatively lowly position as a finance clerk at this

[11] The manuscript is described, and its contents analysed in detail, in Jodogne, 'Recueil', and *Jean Lemaire*, 83, 147–58.

time: the text is carefully written on parchment with gilded *bâtons*, coloured arabesques, initials incorporating watercoloured grotesques, and two small miniatures. The original binding, with bosses and a clasp, survives. There is also a clear (and costly) grammar of legibility: decorated initials and paraphs, in a variety of colours, mark the beginning of texts and stanzas of verse, while headings are usually in red ink. Whatever Lemaire's precise influence over illustrations and layout, this heavy material investment valorizes not only his own poems, but also the literary background he constructs for himself through the anthologized texts.

Secondly, the reference to 'grant sens' reflects the anthology's moral and religious orientation. The texts it transmits form three groups: classical material focusing on the themes of mutable fate, worldly transience, and poetic glory; Latin religious material, largely focusing on the Virgin; and French religious verse, also principally centred on Mary. As Pierre Jodogne has observed, Lemaire is presented as a poet with religious and moral leanings, drawing on a wide literary culture. Yet this statement of Lemaire's priorities may be refined in the light of the note's third noteworthy feature, its versified form. This hints at two aspects of the collection: the stress placed upon poetry's versification and language, and the evident confidence in poetry's capacity to express religious devotion, and to transcend time and hence secure its writers' immortality. Close analysis of the context and physical presentation of the two Lemaire poems reveals the ways in which these tendencies operate.

The first poem, 'Nostre eaige est brief ainsi cõme des fleurs', reflects upon transience from a religious viewpoint.[12] It follows a sequence of Latin texts including a prayer to the Virgin; Ausonius' *Rosae*, a piece then attributed to Virgil, which stresses the brevity of earthly life and beauty;[13] and extracts from Ovid's *Ars amatoria* dealing with poetry and transience. Lemaire's poem not only develops the preceding reflections upon time from a spiritual perspective, but also reveals that Lemaire has been using his own limited time productively: the text constitutes an attempt to obtain poetic renown, and thereby resist the evanescence of worldly things. In this respect, Lemaire practises the sentiment

[12] Fos. 6^v–9^r; edited in Jodogne, 'Recueil', 196–8, from which line numbers are taken.

[13] Lemaire was later to translate this poem: Munn, *Contribution*, 148–52, edits his version.

expressed in the opening verse of the longest Ovid extract: 'Quid petitur sacris, nisi tantum fama, poetis?'[14]

The poem's sophisticated versification reveals Lemaire's technical competence: it employs *rimes batelées*, often highlighted by *virgulae* or *punctūs*. However, Lemaire's virtuosity, and the importance of the manuscript's material form, manifest themselves more strikingly with respect to the piece's allegory. Its initial comparison of earthly life to a rose is superseded by an allegorical analysis of three flowers, or *pensées*, which must be planted in the heart to ensure a virtuous life. These flowers appear in an illustration immediately before the poem, depicting them in blue on a golden shield held by a winged lady (fo. 6^r; Plate 3.1); they are also depicted earlier in the manuscript, in a smaller image of the same shield with the motto 'Penser penser penser. Dire' (fo. 2^r), a line from the poem. The flowers are implicitly compared to the three theological virtues, and the shield blazoned in a more generally moral sense:[15]

> Pallas lassorte/ et prudence le porte.
> Tout sy comporte/ en vray mistere pur
> Le champ est dor/. les pensees dazur.
>
> PAr lor sentent vertu resplendissant.
> Lazur plaisant/. le hault ciel notiffie.
> Nombre de trois/ est tousiours florissant
> Lescu luisant/ noble cueur signifie. (fo. 8^r; cf. vv. 40–6)

An amorous reading is also provided:

> Lescu tousiours denote bien celer
> Lor peu parler/. lazur bien besongner. (fo. 8^v; cf. vv. 60–1)

The miniatures foreground the flower image's polysemy. By juxtaposing both senses of *pensée* ('pansy'/'thought'), the image and motto on fo. 2^r cataphorically highlight the wordplay with which Lemaire's poem, later in the manuscript, underpins the equation of flowers with virtues. The second shield image prolongs the reader's hesitation between alternative readings of *pensée*.

Moreover, the lady in this second image is enigmatic: she has no certain referent in the text. She may represent Prudence, whom the poem describes as carrying the shield. This would underline

[14] *Ars amatoria*, 3. 403; Ovid, *'The Art of Love', and Other Poems*, ed. and tr. J. H. Mozley, rev. G. P. Goold (2nd edn. Cambridge, Mass., 1979), 146. Cf. fo. 3^r.

[15] U. Bergweiler, *Die Allegorie im Werk von Jean Lemaire de Belges* (Geneva, 1976), 107–17, discusses the poem's allegorical senses.

PLATE 3.1 Punningly symbolic flowers: BNF, n. a. fr. 4061, fo. 6^{r} (Cliché Bibliothèque Nationale de France, Paris)

Lemaire's unusual step in linking faith, hope, and charity to the past, present, and future respectively:[16]

> la premiere fleur gente
> Tresdiligente au temps passe reuoit
> La seconde le temps present regente
> Lautre fulgente au futur temps pouruoit. (fo. 7^{v}; cf. vv. 29–32)

The three aspects of time were traditionally associated with Prudence, a virtue entailing memory, intelligence, and foresight.[17] Though it cannot be confirmed, this interpretation is tempting,

[16] Bergweiler, *Allegorie*, 112–14, notes this link's novelty.

[17] See J. Kem, 'Allegorical Images of the Cardinal Virtue Prudence in the Works of the *Rhétoriqueurs*'.

and again accentuates the density with which Lemaire invests the floral allegory.

Visually more unusual is Lemaire's second poem, 'Salut a vous dame de hault paraige', here entitled 'Sensuit vne oroison composee par Iehan le maire de la dicte ville de vallenciennes' (fos. 57^{v}–63^{r}). The *Salve* acrostic is written in red ink, immediately drawing attention to the way in which Lemaire has conjoined French and Latin texts, and to the different patterns which the acrostic forms in each stanza. Furthermore, as Gérard Gros has noted, the positioning of Lemaire's poem is highly significant.[18] It immediately follows Marian poems by Chastelain and Molinet, the former's *Louenge à la trèsglorieuse Vierge* and the latter's 'Quant Terpendreus sa harpe prepara' (fos. 29^{r}–57^{v}).[19] Both these writers lived in Valenciennes, a fact mentioned in the titles which their poems bear in this volume: Lemaire thus presents himself as the third generation in a lineage of Valenciennes poets.[20]

Closer study reveals further implications of the poems' grouping. All three pieces are self-conscious in that they reflect upon poetry and its adequacy as a vehicle for religious praise. Chastelain reflects upon the problems and limitations involved in composing religious poetry:[21]

> Iay fait vng vol/ qui ma nature passe.
> Cuidant actaindre/. a lumiere oultrepasse.
> Dont latempter. plusauant ie delaisse
> Et repassant arriere en terre basse.
> A ma main triste/. et a ma plume lasse
> Repos ioctroye. (fo. 52^{v}; *Œuvres*, viii. 290–1)

Molinet also makes explicit the poet's activity, by signalling the way in which he compares Terpander's harp to the Virgin:[22]

> Ceste harpe/. qui si belle forme a
> Puis figurer. par viues raisons nectes.
> A marie vierge que dieu forma. (fo. 55^{v}; *FD*, ii. 447, vv. 12–14)

[18] G. Gros, 'Sur l'oraison mariale de Valenciennes, ou Jean Lemaire de Belges et ses devanciers'.

[19] G. Chastelain, *Œuvres*, ed. K. de Lettenhove (Brussels, 1863–6), viii. 269–92 (further references will be incorporated into the text); *FD*, ii. 447–9.

[20] Lemaire also follows these authors as part of a larger tradition of historical writing associated with Valenciennes: see G. Small, 'Chroniqueurs et culture historique au bas Moyen Âge'.

[21] See G. Gros, '*Querant l'un œil envers les cieulx estendre . . .* : Étude sur la *Louenge* mariale de George Chastelain', 431–5, 440–4; Cornilliat, *'Or ne mens'*, 519–20.

[22] See Cornilliat, *'Or ne mens'*, 545–55.

Lemaire's poem reflects on religious verse more optimistically than Chastelain's, suggesting that devotional writing is potentially worthwhile and effective:

> Moult est heureu**x e**t de riche auenture.
> Qui par ver**tu** de sa propre nature.
> Peut **se l**euer en contemplation
> Ed**iff**iant par aucune facture.
> **Mil**le dictiers de notable lecture.
> **I**ouxte vos biens. et collaudation. (fo. 59[r])

The positioning of Lemaire's poem after those of his predecessors is also significant in the light of the layout's foregrounding of the *Salve* acrostic. Lemaire's piece does not exhibit as much technical rigour as Chastelain's and Molinet's: its stanzas vary in length (six *onzains*, two *douzains*, and two *quatorzains*, besides the *envoi*), suggesting that Lemaire found it difficult to observe a regular stanzaic pattern when attempting to construct his acrostic. However, the use of red ink draws attention to Lemaire's ingenuity, and thereby implies that his poetic skill is comparable to that of his predecessors. Lemaire has thus inserted himself into a poetic tradition within which he shows his virtuosity to advantage, suggesting he is the rightful successor to Chastelain and Molinet as Burgundian *indiciaire*, and once more staking his claim to renown through poetry.

Lemaire's *livret sommaire*, then, clearly shows his acute awareness of the resources of anthologization, illustration, and layout, which highlight various self-conscious features. The arrangement of texts suggests that Lemaire is defined through his reading and writing, and underlines his confidence in poetry as a means of both expressing moral and religious concerns and obtaining fame; layout focuses attention on his poems' form and language; illustrations help reveal his allegories' dense semantic implications.

The Épître du roi*: Saltÿkov-Shchedrin Library, MS Fr. F. v. XIV, 8*

Though striking in the *livret sommaire*, the visual display of formal and linguistic artfulness is rarer in manuscripts of Lemaire's poetry than in those of Molinet's work.[23] Lemaire's verse, indeed, exhibits less virtuosity of this kind than does Molinet's; it displays self-consciousness more often by foregrounding an intertext than through prosodic intricacy. As previously noted, the *Épître du roi* provides a significant and humorous example of this technique. Lemaire's practice in the *Épître* is conveyed particularly effectively in St Petersburg, Saltȳkov-Shchedrin Library, MS Fr. F. v. XIV, 8, a collection of poetic epistles by various authors, executed for Louis XII or Anne of Brittany no earlier than February 1512.[24] The epistles, some of which are French translations of Latin pieces, relate to Louis's conflicts with Venice and, especially, Julius II, in the period 1509–12; they are all either addressed to or ostensibly sent by Louis. Lemaire's *Épître* immediately follows d'Auton's poem (fos. 66^{v}–81^{r}, 82^{r}–95^{r}), both pieces being explicitly attributed; the juxtaposition makes evident the game of challenge and response, and enables Lemaire's poetic play to be fully appreciated.[25]

However, Lemaire's ingenuity is perceptible not only through a reading of the *Épître* against d'Auton's poem, but also against the codicological horizon of the entire manuscript. The anthologized pieces' political orientation provides a strong context of criticism of Julius, within which Lemaire's attacks on the Pope stand out. Yet the genre of the poems in this volume is an equally significant influence on the reading of Lemaire's text. The verse epistle was currently fashionable in France, thanks partly to Octovien de Saint-Gelais's translation of Ovid's *Heroides*, and partly to the Italian-born court poet Publio Fausto Andrelini, who composed

[23] Layout is, however, used expressively in BNF, ms. lat. 1428. This volume, which largely contains Latin religious material, ends with an untitled Lemaire poem beginning 'A La splendeur de tes haulx raiz sublimes' (fos. 58^{r}–62^{r}; edited in Munn, *Contribution*, 180–3. See also Gros, *Poète marial*, 80–2). The piece incorporates a Latin acrostic, an *anthienne* beginning 'Alma redemptoris mater', highlighted in red ink.

[24] For descriptions, and reproductions of some of the miniatures, see Comte A. de Laborde, *Les Principaux Manuscrits à peintures conservés dans l'ancienne Bibliothèque Impériale Publique de Saint-Pétersbourg* (Paris, 1938), ii. 147–9; G. Tournoy-Thoen, 'Fausto Andrelini et la cour de France', 70–6; F. Avril and N. Reynaud, *Les Manuscrits à peintures en France 1440–1520* (Paris, 1993), 303–5; LeBlanc, *Va lettre va*, 108–9. The manuscript's *terminus a quo* may be established on the basis of Jean d'Auton's *Epistre elegiaque par leglise militãte transmise au roy trescrestien Loys douziesme* (fos. 101^{r}–112^{v}), which includes an account of Gaston de Foix's recapture of Brescia on 18 February 1512 (fos. 107^{v}–109^{r}); I am grateful to Jennifer Britnell for pointing out this detail.

[25] Lemaire's poem also follows d'Auton's in ÖNB, Cod. 2579, fos. 86^{v}–111^{r}, though neither piece is attributed in this volume.

three influential epistles in Latin from Anne of Brittany to Louis in the period 1509–11.[26] Hence this manuscript's grouping of epistles—including Andrelini's three pieces, in French translations by Macé de Villebresme, with the Latin text in the margins—permits readers to compare the differing ways in which these poems modulate and vary the conventions of letter-writing.[27] Certain epistles, for example, stand out by conjuring up strikingly unusual images. Andrelini's first epistle (fos. 1^r–9^v), in which Anne recounts her grief at Louis's absence in Italy and her elation at his victory over Venice at Agnadello in May 1509, includes a vision of Anne fighting alongside Louis:

> Par maintesfoiz me suys deliberee
> Dicy partyr en armes preparee . . .
> Et porter arc comme samiramys. (fo. 4^v)

Still more unusual are d'Auton's *Espitres enuoyees au Roy trescrestien Loys. xij^e. de la les monts par les estatz de france* (fos. 10^r–39^r), also praising Louis after Agnadello. Here, straightforward panegyrics by Eglise and Noblesse are followed by a letter from Labeur, whose references to 'ma femme et ma maignie' (fo. 38^r), and other everyday preoccupations, suggest he is not a personification proper like Eglise and Noblesse, but rather a typical peasant.[28] There follows a comic account of Labeur being visited by a *laquays* returning from the wars, telling a tall tale of his valour at Agnadello ('Le roy meust fait ce iour la cheualier') which, as his host recognizes, 'nest pas sentence deuangille' (fo. 38^v). D'Auton's play upon the conventions of poetic decorum and appropriate style levels reveals how the conventions of verse epistles lend themselves to knowing manipulation, a manipulation which readers will rightly also expect of Lemaire.

Moreover, every piece in the volume is preceded by a miniature, usually depicting the epistle's ostensible sender writing, dictating, or handing over a letter.[29] These images not only further

[26] LeBlanc, *Va lettre va*, 108, 174–8; Tournoy-Thoen, 'Fausto Andrelini'; 'Deux épîtres inédites de Fausto Andrelini et l'auteur du *Iulius Exclusus*'; P. F. Andrelini, *Amores sive Livia*, ed. G. Tournoy-Thoen (Brussels, 1982), 162–6.

[27] On translations of Andrelini, see Andrelini, *Amores sive Livia*, 230–2; on epistolary conventions, see LeBlanc, *Va lettre va*.

[28] J. J. Paxson, *The Poetics of Personification* (Cambridge, 1994), 60–1, distinguishes such 'exemplary property characters' from personifications proper.

[29] The artist is apparently Jean Bourdichon: Avril and Reynaud, *Manuscrits*, 303–5.

thematize the epistolary act, but also present it as public. In most of the illustrations, the writer/sender is watched by other figures: the miniature accompanying Andrelini's first epistle, for instance, portrays Anne writing a letter as a group of women looks on.[30] The act is theatricalized, subjected to the gaze of both the reader/viewer and an internalized audience within the miniatures; epistles may have determinate addressees, but they are not therefore private. The depictions of an audience underline that the texts are not genuine letters, but poetic ones, available to a wide circle of readers.[31] As such, they are open to other writers' responses, to reappropriation, parody, or the continuation of an imaginary correspondence; this is precisely what Lemaire has done in his *Épître*.

The image accompanying Lemaire's *Épître* (fo. 81^{v}) depicts Louis enthroned in the centre of a rich interior, dictating to a young man sitting at his right.[32] To his left stands a messenger, identified by a label as Boreas (mentioned in the *Épître*), while a group of courtiers watches in the background. Royal symbols abound: Louis's throne and canopy are semy of fleurs-de-lis; wall-hangings are semy of porcupines, a standard attribute of Louis;[33] the windows display France's arms, and the monograms L[ouis] and A[nne]. This visual glorification of Louis compensates for the lack of explicit praise in the text, which is constrained by its use of Louis's own voice: the king can hardly fill a letter with self-praise. The miniature is the only one in the volume to depict an epistle being dictated, and since 'maistre Iehan Le maire' is named as the *Épître*'s author (fo. 82^{r}), the viewer is likely to equate the secretary in the image with Lemaire. His position next to Louis is significant, as is the ermine collar he wears, which alludes to Anne's arms. These motifs reflect Lemaire's new allegiance to the French royal court; by January 1512, before the confection of this manuscript, he had become 'Hystoriographe de la royne', as indicated on the title-page of an edition of his *Traicté de la différence des schismes et des conciles* published in that month.[34] While the writing

[30] Fo. 1^{v}; reproduced ibid. 305. [31] See LeBlanc, *Va lettre va*, 4–5.

[32] Reproduced in Laborde, *Principaux Manuscrits*, ii, pl. LXI.

[33] See R. W. Scheller, 'Ensigns of Authority: French Royal Symbolism in the Age of Louis XII', 79.

[34] For further discussion of the *Traicté*, see below, Ch. 4; for this edition, see J. Lemaire de Belges, *Traicté de la différence des schismes et des conciles de l'Église, avec l'Histoire du Prince Sophy' et autres œuvres*, ed. J. Britnell (Geneva, 1997), 60. Further references will be incorporated

figure does not unambiguously represent Lemaire, then, the identification is justifiable and tempting. Consequently, Lemaire enjoys a visual presence in the manuscript, albeit in a subordinate position, noting down Louis's words rather than composing a text himself. The volume thus foregrounds Lemaire's investment in the *Épître* by visually acknowledging his writerly activity, and by providing a context which highlights his manipulation of the intertext.

The Concorde des deux langages*: Carpentras, Bibl. Inguimbertine, ms. 412*

Like the *Épître*, Lemaire's *Concorde des deux langages* appeared in manuscript form before it was printed. Its pre-print state survives in Carpentras, Bibl. Inguimbertine, ms. 412, a well-produced parchment volume dating from 1511, containing two striking illustrations on inserted leaves.[35] These differ from the miniature accompanying the *Épître* in the St Petersburg manuscript, in that they highlight not so much Lemaire's prominence as author as his construction of dense, polysemous allegories and metaphors. It thereby reveals the rhetorical sophistication with which Lemaire has imbued the *Concorde.*

The *Concorde* begins with a prologue in prose, in which the author claims he has been asked to write about an argument over the relative merits of French and Tuscan. He praises both languages, and expresses hopes for peace between France and Florence, which he claims can be found in the Temple of Minerva, but not in the Temple of Venus. The text which follows, he writes, will describe first the Temple of Venus, in *terza rima* 'à la fasson ytallienne', then the Temple of Minerva, in prose and alexandrines, or 'ryme françoise' (6, ll. 98, 110). In the narration proper, the *acteur* sees Venus in a dream, and visits her temple, where Genius officiates, assisted by Dangier, Bel Acueil, and the

into the text. The version of the *Épître* in the St Petersburg manuscript differs in several respects from printed versions, and may therefore represent the poem's initial state; the volume is possibly the presentation manuscript of both the *Épître* and some of the accompanying material.

[35] The manuscript's original owner is unknown. On the *Concorde*'s transmission, see J. Lemaire de Belges, *La Concorde des deux langages*, ed. J. Frappier (Paris, 1947), pp. lxiii–lxvi. Further references to this edition will be incorporated into the text.

Three Graces. Genius delivers a sermon urging French youths to procreate while they are young. The congregation of lovers must donate valuables before being allowed to kiss the relics in the temple; the *acteur* presents a 'petit tableau' (35, l. 144). Though Bel Acueil finds this acceptable, Dangier does not, and the *acteur* is chased away. He wanders into a desert where he finds a large rock, bearing an inscription in alexandrines which explains that those who climb the rock obtain eternal peace. At the summit is the Palace of Honour, which contains the Temple of Minerva. Here, scholars and writers lead a contemplative life, and the French and Tuscan languages are reconciled. To reach the Temple, poets must wait at the bottom of the rock, amid 'estude et labeur et soucy' (42, v. 105), until Honneur admits them. The delighted *acteur* sleeps briefly, but is woken by Labeur Historien, a personification of scholarship. The latter explains that the inscription is the work of Jean de Meun, whom he and the *acteur* compare to Dante, showing the close relationship between French and Tuscan culture; Labeur Historien then proposes a political union between France and Florence. The *acteur* is allowed to remain as Labeur Historien's clerk, and shown the images of the two languages embracing in a magic mirror.

The *Concorde* has suscited much critical debate, primarily over its political significance as a text composed during the Italian wars and addressing the competing claims to cultural precedence of France and Italy. Scholars have discussed, first, whether the political perspectives of the prologue and conclusion extend to the rest of the narrative, which contains no explicit political discourse; and second, whether the temples complement or oppose each other, usually in a political and cultural sense, though sometimes in moral, aesthetic, or philosophical terms.[36] However, the Carpentras manuscript's miniatures evince a reading which

[36] See P. Imbs, 'Jean Lemaire de Belges: *La Concorde des deux langages*'; J. Frappier, 'Nouvelles Réflexions sur *La Concorde des deux langages*'; Jodogne, *Jean Lemaire*, 443–62; F. Rigolot, 'Jean Lemaire de Belges: Concorde ou discorde des deux langages?'; R. Norrell, 'L'Énigme de *La Concorde des deux langages*'; R. Griffin, '*La Concorde des deux langages: Discordia concors*'; D. Fenoaltea, 'Doing It with Mirrors: Architecture and Textual Construction in Jean Lemaire's *La Concorde des deux langages*'; A. Moss, 'Fabulous Narrations in the *Concorde des deux langages* of Jean Lemaire de Belges'; D. Stone, 'Some Observations on the Text and Possible Meanings of Lemaire de Belges' *La Concorde des deux langages*'; H. Campangne, *Mythologie et rhétorique aux XVe et XVIe siècles en France* (Paris, 1996), 102–18; Randall, *Building*, 93–100; Cowling, *Building*, 187–202.

acknowledges principally the libidinal significance of the *Concorde*'s allegory.

The first miniature (fo. 4^{v}; Plate 3.2) depicts the interior of the Temple of Venus, with Venus presiding in judgement, in a position coinciding with the vanishing point. There is no figure identifiable with the *acteur*: viewers must assume the scene is presented through the *acteur*'s eyes. The second image (fo. 20^{v}; Plate 3.3) contrasts sharply with the first: it portrays the Temple of Minerva and Palace of Honour from the outside, and depicts the *acteur* in the foreground.[37]

In the first image, Venus' central position gives her considerable importance. This contrasts with her peripheral position in the text: once the *acteur* enters the Temple, Venus is mentioned only rarely and briefly.[38] Oddly, her seat recalls the chariot in which she first appears: the 'coulons et cignes' which drew her chariot (9, v. 49) are tethered to a kind of platform. This suggests that two scenes, Venus' arrival and the interior of the Temple, have been condensed into one. The miniaturist apparently found this process difficult: the plane on which the birds stand seems, in the context of the rest of the image, to be not horizontal but vertical. Venus' unexpected importance within the Temple results from the integration of the chariot and temple motifs.

The displacement of emphasis onto Venus, and other features of this image, suggest that the artist has recognized the dense sexual metaphors in the description of the Temple, and foregrounded them as the central aspect of the Venus episode. The text contains several clear double entendres, notably an explicit equation of the Temple with a bedroom, followed immediately by a veiled allusion to a vagina:[39]

> Les aultéz sont de litz tres bien paréz,
> Encourtinéz pour eviter spectacle.
>
> Les chappes sont de draps bien figuréz,
> Le propre encens est d'oudeur naturelle,
> Les benoitiers des vaisseaulx corporelz;
>
> Et là dedens jamais eauë ne gelle,
> Se l'aspergès n'est d'estoffe amortie,
> Ou que ung froit vent de craincte ne s'i mesle. (14, vv. 173–80)

[37] Avril and Reynaud, *Manuscrits*, 362, attribute the images to the Lyons artist Guillaume Leroy.

[38] 17, v. 245; 18, v. 272; 20, v. 315; 21–2, vv. 346–51.

[39] See also Cowling, *Building*, 56–7, 193–5.

PLATE 3.2 The Temple of Venus: Carpentras, Bibl. Inguimbertine, ms. 412, fo. 4^v (reproduced by permission of the Bibliothèque Inguimbertine, Carpentras)

PLATE 3.3 The Palace of Honour and Temple of Minerva: Carpentras, Bibl. Inguimbertine, ms. 412, fo. 20v (reproduced by permission of the Bibliothèque Inguimbertine, Carpentras)

Genius' sermon, with its call to procreate, draws attention to such passages' sexual overtones. Moreover, readers are frequently reminded of the *Roman de la Rose*, by the figures of Genius, Dangier, and Bel Acueil, and by Labeur Historien's reference to Jean de Meun. The *Rose* uses religious images with an erotic sense;[40] its prominence as intertext in the *Concorde* thus reinforces the sexual significance of the Temple of Venus. The Temple, then, does not simply symbolize a political scenario, an architectural model, or a particular conception of language or poetry. It does not merely refer to national and cultural debates in the *hors-texte*; it also has a signifying role within the text, as part of a narrative of the *acteur*'s sexual desire.

It is this narrative which the artist emphasizes. By placing Venus at the miniature's vanishing point, where the lines of perspective converge, he makes her the 'back' of the pictorial space. Consequently, Venus is the point towards which the eye of the viewer (both *acteur* and reader) is invited to penetrate. This visual act replicates the act of sexual penetration which is the main preoccupation of the Temple's congregation. Various other visual motifs reinforce this effect. Genius, standing upright, functions as an erect penis pointing towards Venus, appropriately for a figure who embodies procreative power both in the *Concorde* and in a long cultural tradition.[41] Genius' mitre, mentioned in the text (21, v. 331), underlines his phallic quality; *mitre* had the accepted meaning 'glans'.[42] The mandorla-shaped panels on Venus' platform are also erotic in significance. Their shape recalls that of the vagina, and their proximity to Venus' concealed genitalia intensifies the visual suggestion that she invites penetration. Finally, the sword of the figure immediately to Genius' right is clearly phallic, while the prominent purses held by some of the congregation exploit a traditional pun: the purse was a common euphemism for the scrotum.[43] Clearly, the artist has recognized that the *Concorde* is not

[40] See Huot, *'Romance of the Rose'*, esp. 106–9, 148–9, 167–8, 273–4.

[41] See 29–30, vv. 532–3, 550–8. On Genius' generative, phallic properties, see J. C. Nitzsche, *The Genius Figure in Antiquity and the Middle Ages* (New York, 1975), 116–25; G. Raynaud de Lage, '*Natura* et *Genius*, chez Jean de Meung et chez Jean Lemaire de Belges', 140–2; C. Battersby, *Gender and Genius: Towards a Feminist Aesthetics* (London, 1989), ch. 7.

[42] Godefroy, *Dictionnaire*, v. 351.

[43] See G. de Lorris and J. de Meun, *Le Roman de la Rose*, ed. F. Lecoy (Paris, 1965–70), i. 218, v. 7113; iii. 90, v. 19637; A. Tobler and E. Lommatzsch, *Altfranzösisches Wörterbuch* (Berlin and Wiesbaden, 1915–), i. 1080.

simply a political allegory. While the text has several strands of meaning—political, sexual, poetic, philosophical—its sexual elements have been foregrounded.

It is uncertain whether the artist has read the *Concorde* as primarily a narrative of the *acteur*'s desire, a sexual longing frustrated in the Temple of Venus and then sublimated into scholarship under Labeur Historien's tutelage, or whether the Temple of Venus is regarded as representing a libidinous ethos which must be rejected in favour of the Temple of Minerva. However, the strong pictorial contrasts between the two miniatures suggest the latter reading. As I have previously noted, the first illustration depicts an interior, presumably through the *acteur*'s eyes, while the second represents an outdoor scene which includes the *acteur*. Further contrasts can readily be identified. The second picture, for instance, is dominated by muted greens, browns, and blues, in contrast with the first, where much of the décor is golden and many of the congregation wear saturated red robes.[44] Moreover, whereas the Temple of Venus constitutes a symmetrically organized visual space drawing the eye towards the centre, the rock on which the Palace of Honour stands is offset to the right, and Labeur Historien's upward gesture draws the viewer's eye upwards, past the inscription towards the Palace and ultimately to the top of the picture, where the golden rays of the Holy Spirit are visible. These differences in composition thus reflect the different tasks required of the viewer, penetration and elevation; clearly, these activities have strong sexual and spiritual connotations respectively.

Even the buildings are distinguished: the Temple of Venus adopts curvilinear forms, while the Palace of Honour is dominated by spires.[45] Consequently, the Palace appears to be more clearly a Christian church than the Temple of Venus, a difference reflected in the text: while the virtue enshrined in the Palace is not in doubt, the Temple of Venus is characterized by a moral

[44] Verse layout reinforces the visual contrast between the two temples. The tercets of the *terza rima* which describes the Temple of Venus are separated from each other by blank lines, and never allowed to split over two pages, whereas the alexandrines of the inscription form a block of continuous verse.

[45] This contrast is more pronounced than the text's descriptions of the buildings might suggest: see Cowling, *Building*, 192 n. 49.

ambivalence which subverts its magnificence.[46] Indeed, the dangerous ambiguity of the Temple of Venus is suggested pictorially by the problematic horizontal/vertical surface I have noted above. Together with the strong differentiation between the buildings, this implies that the sexuality in the Temple of Venus is criticized, not accepted, by the artist.

Whatever the value ascribed to the Temple's sexual aspect, however, it is undeniably highlighted. In contrast, the images draw little attention to the way in which the *Concorde* thematizes poetry, especially Lemaire's own, which is valorized in competition with Italian and lyric poetry, and which displays its ingenuity by appropriating and reworking motifs from the *Roman de la Rose*.[47] The *Rose* allegories in the first image, and the 'French' alexandrines in the second, recall this aspect of the text. Similarly, the depiction of the *acteur* below the morally impeccable Palace of Honour, but not inside the ambiguous Temple of Venus, may suggest that Lemaire is being visually associated with a tradition of French didactic poetry rather than Italian amorous verse. But the principal expression of these themes is in the prologue, where the narrator (who seems to coincide with both the *acteur* and Lemaire himself) switches attention from a linguistic debate to his own use of poetic forms. He thus encourages a reading of the *Concorde* as focusing on literary creation.

This means of highlighting Lemaire's writerly investment in his text complements the effects of the miniatures, which show the sophistication of the allegorical network Lemaire constructs. His allegories cannot be fully appreciated without considering their relationships to other metaphors: the Temple of Venus, for example, partly relies upon the use of figures from the *Roman de la Rose*. The miniatures underline this tendency by establishing further metaphorical relationships, absent from the text itself, such as the sexually suggestive purses and panels in the Temple of Venus. While the *Concorde*'s political and cultural implications cannot be denied, Lemaire's activity in elaborating a densely worked signifying structure is made prominent.

[46] Moss, 'Fabulous Narrations', 22–6. Cowling, *Building*, 195–8, notes literary precedents for the negative valorization of eroticized buildings.

[47] See Cowling, *Building*, 201–2; C. J. Brown, 'Jean Lemaire's *La Concorde des deux langages*: The Merging of Politics, Language, and Poetry'.

The Couronne margaritique*: ÖNB, Cod. 3441*

The variety of ways in which manuscript illustrations reflect Lemaire's different self-conscious techniques is already apparent. In the St Petersburg manuscript, the *Épître*'s accompanying miniature foregrounds Lemaire's prominence as author and new allegiance to the French court; in the Carpentras volume, illustrations accentuate his use of complex allegory. To complete this survey of the diverse modes of text–image interaction in Lemaire manuscripts, I now turn to what is probably the presentation manuscript of the *Couronne margaritique*, ÖNB, Cod. 3441. As previously noted, the *Couronne*'s prologue thematizes poetry, and, by alluding to Lemaire's previous commemorative compositions, encourages identification of the first-person prefatory voice with Lemaire as historical author. However, Lemaire's writerly investment in the *Couronne* is also highlighted in the body of the narrative, which exhibits a strongly reflexive quality. The manuscript's illustrations draw attention to this quality, as well as to other self-conscious features of the *Couronne.*

A brief account of Lemaire's text permits a proper appreciation of its reflexivity. The narrative begins by recounting Philibert of Savoy's murder by the figures of Infortune and Atropos; Infortune, in disguise, gives Philibert cold water to drink after a strenuous hunting expedition, while Atropos shoots him with her bow. On Philibert's death, his wife Margaret of Austria despairs, while Hebe laments his loss. Vertu, however, sends Prudence and Force to console Margaret. Infortune is defeated when Margaret recovers her spirits, and throws himself from a tower. Vertu decides to reward Margaret by constructing her a crown. She chooses ten personified virtues, each beginning with a letter of the name *Marguerite*; pairs each virtue with a gemstone beginning with the same letter; and arranges them into a ring. Vertu then summons ten great scholars and writers to explain why the virtues and stones are appropriate for Margaret. In erudite passages taking up most of the *Couronne*, each scholar shows that a virtue and its accompanying stone represent qualities which Margaret displays; in each case, Margaret is also compared to another famous lady whose name begins with an appropriate letter, and who exhibited the virtue in question. Meanwhile, Varro's daughter Martia paints the circle of virtues, producing a design from which Merite will be able to construct the crown out of precious materials; the *Couronne* ends at this point, with the crown not yet completed.[48]

It is in the scholars' speeches that the *Couronne*'s reflexivity is apparent. The scholars must justify the crown's metaphorical appropriateness for Margaret, as Vertu explains to them:

> nous voulons que vous nous sachiez à dire, si linuention comprinse en nostre imaginatiue, est bien conduite et mise à effect selon la raison, et sil y ha point de discrepance quant à lobiect pretendu. Cestadire si les excellences de ces dix Vertus, et les proprietez des dix gemmes, ont concordance si mutuelle, que delles puist resulter, comme en vn miroir trescertain, le vif exemplaire de la dame dessus mentionnee. (*Œ*, iv. 58)

The final speaker, Vincent of Beauvais, proclaims their success in this task: 'notre Princesse Marguerite ainsi que son nom seul comprend toutes les premieres lettres des noms des Dames dessusdites, aussi elle seule en tous ces cas, tant de felicité que daduersité, ha suffisamment fourny à toutes lesdites vertueuses labeurs' (*Œ*, iv. 153). Yet, in justifying Vertu's project of constructing a crown to reward Margaret, the scholars simultaneously justify Lemaire's project of constructing a literary text in praise of Margaret, centred on the crown image. The crown is motivated on two levels: explicitly and intradiegetically, as an object designed by Vertu; implicitly and extradiegetically, as an image conceived by Lemaire. The scholars, then, explain the relevance to Margaret of both Vertu's crown and Lemaire's text, which are in a sense analogous. Further analogies between crown and text readily suggest themselves. Both commemorate Margaret's virtues; both must be read (the crown must be interpreted by the scholars). Moreover, both are sumptuously constructed. The *Couronne* comprises sections of both prose and verse, the latter in varying stanzaic forms; the scholars' speeches in French incorporate citations in Latin; there are numerous speakers (narrator, allegories, and scholars); different interpretations are applied to the several elements of the crown (exempla of feminine virtue, gemstones, virtues) with encyclopaedic erudition.[49] This variety of literary

[48] Jodogne, *Jean Lemaire*, 215, observes that a second book of the *Couronne* was apparently composed, but is now lost. See also Randall, *Building*, 76–85.

[49] Jodogne, *Jean Lemaire*, 225–48, discusses the *Couronne*'s encyclopaedic qualities; see also Randall, *Building*, 80–1.

forms and discourses, which Lemaire synthesizes into an impressive whole, echoes the diversity and richness of the materials from which the crown is to be produced. It is these strong affinities between Vertu's crown and Lemaire's poem which render the text reflexive in the scholars' speeches. By justifying the crown image on which it is based, and implicitly proving Lemaire's ingenuity in elaborating such an appropriate allegory, the poem speaks of itself.

The Vienna manuscript is of particular interest, for various reasons. It is probably the *Couronne*'s presentation copy, and was perhaps produced under Lemaire's supervision. A note on fo. 1^{r} mentions that Philip the Handsome received the volume, and apparently confirmed that Lemaire would succeed Molinet as Burgundian *indiciaire*, on 6 June 1505;[50] the manuscript seems to have passed after Philip's death to his sister Margaret, who had it bound in black velvet, presumably in memory of Philibert.[51] Though not on parchment, the volume contains nine watercolour paintings of very high quality. A contemporary hand has amended the text at various points, suggesting a process of authorial revision. The *Couronne*'s title explicitly attributes it to Lemaire, and advertises his prestigious position at Margaret's court: 'La Couronne margariticque [ou aut'ment le triũphe dhonne^{r}/.] Composee par Ieh' lemaire Indiciaire et historiographe/. de madame marguerite daustriche et de bourgoigne/. Duchesse de sauoie/. dame de bresse ⁊ c̃' (fo. 2^{r}; words in square brackets have been added in the margin). A further introductory note provides additional details of the text's genesis, and of Lemaire's career, by mentioning that the *Couronne* was 'cõmence en la ville dannessy en sauoie lan mil cincq cens et quatre/. ou mois de septembre Et acheue (ou chasteau du pont dains ledit an en mars/)' (fo. 1^{v}; words in parentheses have been deleted).

The grammar of legibility is unsophisticated, but clear. Plain red initials mark sections of verse and paragraphs of prose. The beginning of the prose narrative is adorned by a larger initial, decorated with motifs proclaiming Lemaire's allegiance: an impe-

50 Jodogne, *Jean Lemaire*, 85.

51 O. Pächt and D. Thoss, *Die illuminierten Handschriften und Inkunabeln der Österreichischen Nationalbibliothek: Französische Schule II* (Vienna, 1977), ii. 87–91, and M. Debae, *La Librairie de Marguerite d'Autriche* (Brussels, 1987), 49–53, describe the manuscript and outline its history.

rial eagle, the *briquets* of the Order of the Golden Fleece, and golden daisies (*marguerites*). Headings, and the Latin citations in each scholar's speech, are underlined in red. This modest visual differentiation between textual elements' functions and languages reflects the *Couronne*'s formal heterogeneity, drawing attention to Lemaire's compositional achievement.

However, the manuscript's most important physical feature is clearly its miniatures. These have a partly punctuative role: all but the first are preceded by at least one blank page, and situated between a prose and a verse section. Consequently, they further underline the text's formal variety. However, they also relate in important ways to various self-conscious aspects of the *Couronne*. The opening miniature (fo. 2^{v}; see cover illustration) provides a striking instance. It faces the prologue, whose presentation of Lemaire it renders still more suggestive. The poet is depicted writing at his desk, but it is uncertain whether the scene is an indoor or an outdoor one. The writer seems to be indoors, but his room apparently opens out onto a landscape, perhaps representing what he is imaginatively seeing. On this reading, the miniature clearly foregrounds Lemaire's creative activity, complementing the prologue's thematization of his work. Lemaire's investment in the text is all the more strongly accentuated since what the poet visualizes is not a coherent narrative scene; rather, he sees diverse motifs and events, such as Philibert riding, and a grieving Margaret and her retinue in the background, which he will ultimately order into a narrative.

Furthermore, the poet's black clothing reflects the prologue's references to funereal writing and to the 'papier noir' (fo. 3^{v}) which Lemaire must use. While the arms of Philibert and Margaret on the left-hand wall show Lemaire's allegiance, and the clock beside them suggests the theme of mortality, the birdcage below the arms may allude to Lemaire's *Première Épître de l'Amant vert*, written in the voice of Margaret's recently deceased pet parrot, upon which he was working at the same time as the *Couronne*.[52] This reminds the reader that Lemaire has other works to his credit; the figure in the

[52] Jodogne, *Jean Lemaire*, 254, dates the *Première Épître* to May 1505. The note on fo. 1^{v} of the Vienna manuscript claims the *Couronne* was completed in March 1505, but that on fo. 1^{r} suggests the volume was not presented until June. The intervening months may well have been spent in making the amendments (ibid. 215), and in painting the miniatures. It is, therefore, highly probable that the composition of the *Couronne* and that of the *Première Épître* overlapped.

miniature is not simply a passive narrator, but a creator, whose work extends beyond the *Couronne* into other texts. The books depicted on the author's table and above his head reinforce this suggestion, as well as underlining Lemaire's erudition. Hence the miniature tempts the viewer to equate the writer it depicts with Lemaire the historical poet, a conflation similar to that encouraged by the prologue.

Most of the miniatures illustrate episodes from the narrative of Philibert's death and Margaret's grief. Two images, however, represent the crown motif. The first depicts Vertu showing the virtues, standing in a circle, to the scholars; Savoir Humain opens the door to the scholars, while Martia paints the virtues (fo. 40^{v}; Plate 3.4). The second portrays the crown itself (fo. 116^{v}; Plate 3.5). These miniatures strikingly reinforce the crown–text equivalence, and extend it to embrace the manuscript.

The image of Vertu and the circle of virtues precedes the text's account of this scene, so that the viewer is ignorant of the circle's significance; at this point, he knows only that Vertu has decided to reward Margaret with an 'aureolle triumphalle et permanente pour tesmoignaige perpetuel de son labeur' (fo. 31^{v}; *Œ*, iv. 45). Hence the viewer/reader is in the same position as the scholars. Like them, he is presented with an enigmatic scene of which he must make sense; like them, he is enlightened by Vertu:

> Prenez doncq toutes les lrẽs capitalles de leurs [sc. the virtues'] noms/. en cõmencant par celle qui porte vne M. en chief/. Et voyez se iceulx dix caracteres indiuiduz de lalphabet/. scauroient former le nom de la plus vertueuse et plus fortunee p^{i}ncesse qui soit aujourduy viuant sur terre/. Et dabondant donnez vous garde/ de ces dix pierres precieuses qui sont es frontz de ces dix pucelles/. Et vous trouuerez/ par la composition de leurs premieres lettres semblable signiffiance. (fo. 43^{v}; *Œ*, iv. 58)

The scholars' interpretative activity, represented in both narrative and illustration, thus mirrors the reader's activity outside the narrative and pictorial space. As the scholars attempt to read the circle properly, the reader similarly attempts to interpret not only Lemaire's text, but also the miniature. This reinforces the equivalence between the crown (or its basis, the circle of virtues) and the poem: both require careful elucidation. Moreover, the illustrated manuscript is analogous to both crown and text: the miniatures, too, demand study and interpretation.

PLATE 3.4 The circle of virtues: ÖNB, Cod. 3441, fo. 40^{v} (Photo: Bildarchiv/ÖNB Wien)

PLATE 3.5 Margaret's crown: ÖNB, Cod. 3441, fo. 116v (Photo: Bildarchiv/ÖNB Wien)

The final miniature underlines this threefold analogy between crown, poem, and manuscript. It is uncertain whether this image depicts the crown itself, or Martia's painting of it. The latter interpretation is eminently defensible, for only the painting, not the crown, has been completed at the end of the text. If the miniature is regarded as portraying Martia's picture, however, the viewer is strongly tempted to replace a relationship of mimesis with one of identity, to see the miniature not as depicting but as being Martia's painting. The image thus potentially conflates a narrated object (Martia's painting) with a real physical object (the painted page). As a result, it exemplifies the propensities which characterize the *Couronne* as transmitted in this manuscript: the equation of imaginary objects, Lemaire's fictional text, and the book. Furthermore, these final illustrations draw attention to the *Couronne*'s reflexivity: the more prominent the crown/text analogy, the clearer it becomes that the scholars' speeches relate to Lemaire's text as well as Vertu's crown.

The equivalence between crown, text, and manuscript may be further explored through Susan Stewart's notion that the souvenir is an object which initiates a narrative. According to Stewart, souvenirs inherently evoke nostalgic accounts of their owners' past experience.[53] The crown functions in precisely this fashion: the scholars who interpret it not only explain the relevance of the virtues and gemstones, but also relate biographical anecdotes about Margaret. Hence the crown commemorates Margaret's past life, in the same way as Lemaire's poem. Moreover, the Vienna manuscript was itself a souvenir: owned by Margaret and bound in black velvet, it commemorated both herself and Philibert. Crown, narrative, and manuscript thus all operate as souvenirs. Again, the reflexive analogy between crown and text is underlined, and extended to include the manuscript.

Conclusions

The Vienna manuscript, then, illustrates a wide variety of the *Couronne*'s self-conscious features, drawing attention to both Lemaire's strikingly personal prologue and the scholars' implicit

[53] S. Stewart, *On Longing: Narratives of the Miniature, the Gigantic, the Souvenir, the Collection* (Baltimore, 1984), 136.

thematization of the text's governing crown metaphor. Comparison with the other volumes studied reveals that Lemaire's manuscripts accentuate a strikingly diverse array of literary techniques. Lemaire's *livret sommaire* is carefully anthologized to define his poetic values, and displays his use of language and metaphor to advantage through layout and illustration; the St Petersburg manuscript highlights his ingenious play on d'Auton's poem, and visually flatters Louis XII; the Carpentras volume's illustrations respond expressively to the *Concorde*'s dense metaphorical system. As I shall now show, this variety contrasts with the rather more selective presentation of Lemaire's works in print.

4

EDITIONS OF LEMAIRE'S POETRY

Introduction

Many editions of Lemaire's poetry were published with his *Illustrations de Gaule et singularités de Troie*, a major historiographical work in three volumes, printed frequently between 1511 and 1549.[1] It is important to grasp this work's content and political orientation at the outset, since the *Illustrations* and the material published alongside it each influences the ways in which the other may be read. A learned text, the *Illustrations* incorporates epic and romance elements, and may be interpreted morally; however, its principal concern is to retrace, through Troy and ultimately to Noah, the genealogy of the inhabitants of 'Gaul', and to assert this region's cultural pre-eminence in Europe.[2] 'Gaul' is an ambivalent term: it often denotes an area covering both France and the Austro-Burgundian lands (whose praise thus reflects Lemaire's desire for European unity in the face of Turkish threats), but elsewhere corresponds to France.[3] These inconsistencies result primarily from Lemaire's changing personal circumstances. His move from Margaret's to Anne's court represented a significant transfer of allegiance in itself; it was accompanied by a period of uncertainty, during which he had not yet entirely broken with Margaret nor become established at the French royal court; moreover, after the move, Lemaire remained in contact with Margaret.[4] In the light of his shifting, equivocal position, Lemaire revised the *Illustrations*, which had originally glorified above all the Austro-Burgundian

[1] *Œ*, i–ii. J. Abélard, *'Les Illustrations de Gaule et singularitez de Troye' de Jean Lemaire de Belges: Étude des éditions—genèse de l'œuvre* (Geneva, 1976) describes the *Illustrations*' editions.

[2] See Jodogne, *Jean Lemaire*, 404–39; Abélard, *'Les Illustrations'*; W. Stephens, *Giants in Those Days: Folklore, Ancient History, and Nationalism* (Lincoln, Nebr., 1989), 139–84; M. Rothstein, 'Jean Lemaire de Belges' *Illustrations de Gaule et singularitez de Troyes* [*sic*]: Politics and Unity'; J. Kem, *Jean Lemaire de Belges's 'Les Illustrations de Gaule et singularitez de Troye': The Trojan Legend in the Late Middle Ages and Early Renaissance* (New York, 1994).

[3] See J. Abélard, 'Les *Illustrations de Gaule* de Jean Lemaire de Belges: Quelle Gaule? Quelle France? Quelle nation?', to which the following discussion is indebted.

[4] See Jodogne, *Jean Lemaire*, 127–9.

lineage of 'Gaul' in the broader sense, to give 'Gaul' an occasionally narrower meaning. He thereby rendered his text more appealing to a French audience, ultimately pleasing both Margaret and Anne.

This, then, is the context within which much of Lemaire's poetry was printed. His *Épîtres de l'Amant vert* were usually printed with book I of the *Illustrations*.[5] Several other texts appeared in three small collections, named after the first pieces in each. The *Épître du roi* introduces a group of occasional poems relating to the Italian wars, as discussed below. The *Légende des Vénitiens* (Œ, iii. 361–407) accumulates historical proofs of Venetian depravity, justifying the war waged against Venice by an alliance, including France, created in the wake of the Peace of Cambrai.[6] Its first edition, also including two occasional poems, the *Plainte du Désiré* and *Regretz de la dame infortunée*, appeared in 1509;[7] subsequent printings accompanied the *Illustrations*. Finally, the *Traicté de la différence des schismes et des conciles* gives its name to an anthology contributing to a French propaganda campaign in favour of Louis XII, who had called for a General Council of the Church to condemn Julius II's attempts to drive the French from Italy.[8] These collections were published in configurations reflecting the *Illustrations*' differing modes of publication.[9] Early editions, like the three books of the *Illustrations*, appeared as separate volumes, to be bound into composite collections by the bookseller; after 1524, the texts were usually published within single-volume collective editions, though the *Légende* was not printed between 1516 and 1549.

Lemaire was probably involved in publishing the *Illustrations*' early editions, whose privileges are granted to Lemaire himself rather than to a printer.[10] These editions' paratext reflects the ways in which Lemaire promoted both his material interests and his politically conciliatory attitudes through the printed book. In the

[5] J. Lemaire de Belges, *Les Épîtres de l'Amant vert*, ed. J. Frappier (Geneva, 1948); see Jodogne, *Jean Lemaire*, 254–68.

[6] See Jodogne, *Jean Lemaire*, 320–43; Brown, *Shaping*, 58–63. A critical text, edited by Anne Schoysman, is forthcoming.

[7] Brown, *Poets, Patrons, and Printers*, 49–58, discusses this edition.

[8] See Lemaire, *Traicté*, 11–47; Jodogne, *Jean Lemaire*, 344–86; J. Britnell, 'The Antipapalism of Jean Lemaire de Belges' *Traité de la difference des schismes et des conciles*'.

[9] See Abélard, '*Les Illustrations*', esp. 35–49, 131–2.

[10] On the *Illustrations*' privileges, and other features discussed in this paragraph, see ibid. 23–9, 59–74, 77. An earlier example of Lemaire's involvement in publication is the *Légende*'s first edition: see Brown, *Poets, Patrons, and Printers*, 49–53.

earliest editions, published during the period of uncertainty, liminary texts address both French and Burgundian figures, and woodcuts glorify Margaret as well as Louis and Anne: Lemaire's potential audience is thus maximized. Once Lemaire was secure in Anne's employment, the woodcuts' political orientation swings decisively towards France (simplifying the *Illustrations*' stance in the process). Moreover, among the liminary material are letters to and from various writers of Lemaire's acquaintance. These promote Lemaire's interests by praising his work, thereby helping to guarantee his scholarship for the reader. In addition, they set the author in a network of real-life relationships, showing that he has an identity in the world outside the text, and ultimately encouraging readers to conflate Lemaire's first-person voice within the *Illustrations* with Lemaire as historical author.

Such self-conscious techniques are also often evinced by the *mise en livre* of other works published with the *Illustrations*. Yet, while the early *Illustrations* editions are important for an understanding of Lemaire's relationship to printing, they cannot constitute the sole frame of reference within which to examine editions of his poetry. The poems have their own distinctive self-conscious features, which editions foreground to varying degrees; these must be addressed whenever relevant. Furthermore, texts' presentation in print cannot always be plausibly ascribed to Lemaire's own priorities in the same way as that in the early *Illustrations* editions. Some pieces were printed without Lemaire's consent, while Lemaire had no influence over his works' publication after 1516 at the latest.[11] Hence, while some editions apparently form part of Lemaire's project, others must be treated as indices of his work's reception.

In this chapter, I examine the presentational features of three groups of Lemaire editions, which exhibit significant or representative tendencies in his printed output. First, I consider the *Temple d'Honneur et de Vertus*, the first of Lemaire's major texts to be printed, in 1504; this displays both a fascinating variety of self-conscious literary techniques, and significant presentational changes over some forty-five years. I then address the editions published in the Burgundian Netherlands between 1507 and 1509, which transmit texts of strongly pro-Burgundian character.

[11] See J. Britnell, 'La Mort de Jean Lemaire de Belges, l'édition de 1517 du *Traité des schismes et des conciles*, et les impertinences d'un éditeur'.

Finally, I discuss the *Épître du roi* volume, the most varied and, in respect of *mise en livre*, the most interesting of the collections published alongside the *Illustrations*.

Le Temple d'Honneur et de Vertus

The *Temple d'Honneur et de Vertus* recounts, in allegorical form, the death and apotheosis of Pierre II of Bourbon, Lemaire's former employer.[12] Lemaire originally presented the text to Louis of Luxembourg, whom he was apparently already serving before Pierre's death; however, Louis died soon afterwards, and Lemaire rededicated the *Temple* to Pierre's widow, Anne of France. Editions of the *Temple* contain both dedications, though Lemaire stated in his dedication to Anne that he had presented the text to Louis 'peu de jours avant son trespas à Lyon',[13] implying the first redaction was in a manuscript now lost.

The dedication to Anne in the first edition, published by Anthoine Vérard early in 1504,[14] suggests that the *Temple*'s printing and rededication were simultaneous. Moreover, it implies that Lemaire and Vérard collaborated in producing the edition:[15]

> puis que le livre (tel qu'il est) m'est demouré entre les mains [after Louis's death], après l'avoir communicqué à Anthoine Verard, libraire de Paris, lequel l'a bien voulu mettre sur ses formes impressoires et le publier par tout . . . , je vous en fais ung petit present, et le dedie et intitule à vostre haultesse tresexcellente. (45)

The printed redaction also includes a laudatory verse epistle by the Paris-based poet Guillaume Cretin; Lemaire's decision to incorporate this piece may explain why the *Temple* was first printed in Paris, rather than Lyons, a milieu to which it is strongly linked.[16] Soon after the publication of Vérard's edition, Michel Le Noir

[12] See Jodogne, *Jean Lemaire*, 170–203; J. B. Wadsworth, 'Jean Lemaire de Belges and the Death of Pierre de Bourbon', and *Lyons 1473–1503: The Beginnings of Cosmopolitanism* (Cambridge, Mass., 1962), 177–84; Cowling, *Building*, 172–87. Brown, *Poets, Patrons, and Printers*, 38–49, discusses aspects of the *Temple*'s early editions.

[13] J. Lemaire de Belges, *Le Temple d'Honneur et de Vertus*, ed. H. Hornik (Geneva, 1957), 44. Further references will be incorporated into the text.

[14] Paris, Bibl. Mazarine, Rés. 21716.

[15] See Brown, *Poets, Patrons, and Printers*, 41; Winn, *Anthoine Vérard*, 73–5, 194–7.

[16] Wadsworth, *Lyons*, 178–9.

printed an unauthorized copy of the *Temple*.[17] Subsequent editions appeared at long intervals: the next was probably printed by Le Noir's son Philippe,[18] and a fourth by Denis Janot and Alain Lotrian.[19] The final sixteenth-century edition, printed by Jean Réal for Jean Bonfons in 1548–9, formed part of a collective volume including the *Illustrations*, the *Épître* collection, and the *Traicté* and associated material, immediately after which the *Temple* appears.[20]

The *Temple* has several noteworthy self-conscious aspects. Most immediately striking is its formal variety. The prose dedications are followed by the songs of seven shepherds with Virgilian names, representing provinces ruled by Pierre and Anne, or Pan and Aurora in the text's pastoral allegory; these hymns praising Pierre are in alternating decasyllabic *huitains* and pentasyllabic *dizains*. The narrator and some of the shepherds then recount and comment on Pierre's death, in *terza rima*. Aurora subsequently dreams that she and the shepherds are transported to the eponymous Temple, where they meet six personified virtues whose initial letters (embroidered on their robes) make up the name *Pierre*. Aurora/Anne is consoled by the figure of Entendement, who narrates Pierre's reception in the Temple, where he is crowned as one worthy of posthumous renown. Anne and her daughter are led into the temple, while the shepherds carve epitaphs for Pierre on its exterior walls; Entendement proclaims Pierre's death, and in a *double ballade* exhorts all to follow Pierre's virtuous example. The dream section is largely in prose, though the epitaphs and certain other speeches are in verse of various forms. In its formal diversity, the *Temple* has the quality of a rhetorical *tour de force*.

Lemaire's formal innovations reinforce the text's intricacy. His use of *terza rima*, for instance, is the first in an original work in French.[21] Anne's dream also departs from the typical allegorical

[17] Paris, Michel Le Noir, 6 April 1504 (BNF, Rés. Ye 846). An undated state also exists; BNF, Rés. Ye 219. See Brown, *Poets, Patrons, and Printers*, 48–9.

[18] n.d.o.p. [Paris, Philippe Le Noir, *c.*1520] (BNF, Rés. Ye 859).

[19] Paris, Alain Lotrian and Denis Janot, n.d. [1530–4] (BL, C. 39. b. 14). The approximate date covers the period of the Janot–Lotrian partnership: S. P. J. Rawles, 'Denis Janot, Parisian Printer and Bookseller (*fl.* 1529–1544): A Bibliographical Study', Ph.D. thesis (Warwick, 1976), i. 18–20.

[20] BNF, Rés. La^2 12; see Abélard, '*Les Illustrations*', 153–5.

[21] Jodogne, *Jean Lemaire*, 190.

dream in *rhétoriqueur* poetry, which is dreamt by the first-person narrator, and frames the whole narrative. Here, the dreamer is not the narrator, and the dream does not begin until almost halfway through the piece.[22] Moreover, its end is not recorded: once Anne enters the Temple, the viewpoint switches to the shepherds and Entendement.

The *Temple*, then, displays a self-conscious attention to prosody, and to the tradition of allegorical poetry off which it plays. Its use of pastoral allegory is similarly knowing: while the pastoral genre had been used for political or allegorical poetry since Virgil,[23] Lemaire here exploits a particular intertext, Molinet's *Bergier sans solas* (*FD*, i. 209–24). The *Temple* shares certain prosodic features with this piece, but its greater formal variety, and plurality of speaking voices (the shepherds, the narrator, Entendement, the six virtues), render it more sophisticated.[24] As often in his early works, Lemaire proves his compositional skill by surpassing the acknowledged master, Molinet.

The *Temple*'s most strikingly self-conscious feature, however, is its use of *mise en abyme*.[25] Lemaire's description of the Temple replicates the text which contains it: the building is 'ung ediffice sumptueux à merveilles, à maniere d'ung temple anticque en ouvraige mais riche oultremesure en sa façon, lequel donnoit de prime face esbahyssement à l'oeil' (74, ll. 677–80). Lemaire's poem is itself a monument to a dead luminary using the 'anticque' genre of Virgilian pastoral, a work of dazzling density which challenges readers' interpretative competence, just as the letters on the virtues' robes baffle the shepherds: 'L'ung disoit que la premiere qui avoit les PP estoit Pales, la bonne deesse des pasturaiges. L'autre affermoit que c'estoit Pallas, deesse de prudence' (75, ll. 716–19). Furthermore, the *Temple* engenders and perpetuates further poetry, as the shepherds add their epitaphs to it. This expresses Lemaire's position with regard to the pastoral genre: Molinet, and other poets stretching back to Virgil, constitute the edifice of poetic tradition which Lemaire assimilates and elabo-

[22] The passage recounting the dream begins at l. 641 in Hornik's edition of the *Temple*, which numbers 1,417 lines in all. On the dreamer, see Bergweiler, *Allegorie*, 124–5.

[23] A. Hulubei, *L'Églogue en France au XVIe siècle: Époque des Valois (1515–1589)* (Paris, 1938), ch. 2.

[24] See Jodogne, *Jean Lemaire*, 174.

[25] Cowling, *Building*, 176–87, judiciously analyses the text–building equivalence, especially the Temple's polysemy, in terms which the following discussion complements.

rates upon by adding his own text to the existing corpus. Hence Lemaire's shepherds represent both readers, who must make sense of the poetic monument before them, and Lemaire himself, who appropriates and adds to the monuments of past literature. By metaphorically thematizing Lemaire's activity, this *mise en abyme* confirms the *Temple*'s highly self-conscious qualities.

The editions foreground some of these qualities in particular ways. Vérard's title-page, like those of all editions except the last (which has no separate title-page), both names Lemaire and inserts him into Molinet's poetic lineage: 'Le temple dhoneu[r]. et de vertus. Compose par iehan le maire disciple de Molinet/ A lhonneur de feu monseigneur de bourbon.' In his dedication to Louis, Lemaire reinforces this connection between himself and Molinet, by referring to himself as a 'trespetit et incongneu disciple et loingtain imitateur, desirant suyvre les vestiges mesmement de monseigneur l'indiciaire archiducal, maistre Jehan Molinet, mon precepteur & parent' (48). The references to Molinet in this passage indirectly evoke specific Molinet poems which recall the *Temple*. These include not only the *Bergier sans solas*, but also the *Chappellet des dames* and *Trosne d'Honneur* (*FD*, i. 100–26, 36–58), which both, like Lemaire's text, contain groups of virtues whose initial letters make up a luminary's name. Against this intertext, the *Temple*'s innovative aspects may be more fully appreciated: due to its use of *terza rima* and unusual allegorical dream, it compares favourably with similar pieces by Molinet.

Liminary material also employs rhetorical techniques which resemble and therefore highlight those of the *Temple*. The dedication to Louis begins with a brief account of recent events, narrated in pairs: the deaths and accessions of two popes, the deaths of two French dukes, the burning of two buildings (47–8). This parallels the distribution of the six figures who greet Pierre at the entrance to the Temple: 'deux y en avoit qui portoient habit de majesté royalle, deux d'auctorité ducalle, et deux de dignité pontificalle' (85, ll. 988–90). More significantly, the dedications echo the *mise en abyme* motif. Lemaire tells Louis that writers have plenty of material 'pour s'employer à graver en marbre, c'est-à-dire, en perpetuelle cronicque, les merveilleux cas funebres qui adviennent de jour en jour' (48), while in the dedication to Anne he refers to his poem as 'ung petit ediffice quant à la structure mais grant quant au subgect dont il est fondé' (44).

Besides foregrounding Lemaire's compositional virtuosity, the liminary pieces help valorize his work for the reader, as in the *Illustrations*. This is especially clear in Cretin's epistle, which furnishes Lemaire with the recommendation of an illustrious contemporary.[26] A more indirect form of self-promotion operates in the dedications, which multiply references to real historical details. Their headings enumerate the addressees' titles: Anne, for instance, is referred to as 'duchesse de Bourbonnois et d'Auvergne, contesse de Clermont, de Forestz, de Gyen, et de la Marche, dame de Beaujeu, et cet.' (43). Within both dedications are numerous references to historical figures, and autobiographical allusions: Lemaire informs Anne that he had previously presented the text to Louis, and calls himself 'je, Jehan le Maire, natif de Haynault' (44), while in the dedication to Louis he claims that the painter Jean Perréal had encouraged him to offer Louis his poem (49). Such elements do not only call attention to Lemaire's existence outside the text; they also impress the reader, by revealing the illustrious circles in which Lemaire moves. Hence, as Cynthia Brown has observed, the titular and liminary material takes advantage of printing to promote Lemaire's talents to a wide public; Lemaire's awareness of the link between printing and poetic renown is acute.[27]

However, liminary material is not the only expressive presentational feature of the *Temple* editions. Layout also underlines aspects of the text's form and allegory. For instance, in the two Le Noir editions and the Réal/Bonfons volume, whose pages are large enough to permit it, the pentasyllabic *dizains* of the opening pastoral section are presented in two parallel columns (Plate 4.1). Though the layout is presumably an economy measure, saving space on the page, it institutes a striking visual contrast between these stanzas and the decasyllabic *huitains*, presented in a single column. The variety of Lemaire's metres is thereby highlighted. Moreover, readers may initially be tempted to read the *dizains* as decasyllabic *cinquains*, split into two columns of half-lines, and only realize that the columns are to be read sequentially when this initial reading fails to make sense. In this case, the layout accentuates the verse's dense sound patterning (there are enough rhymes to make either direction of reading look initially

[26] Cowling, *Building*, 173; Brown, *Poets, Patrons, and Printers*, 42.

[27] Brown, *Poets, Patrons, and Printers*, 42–7.

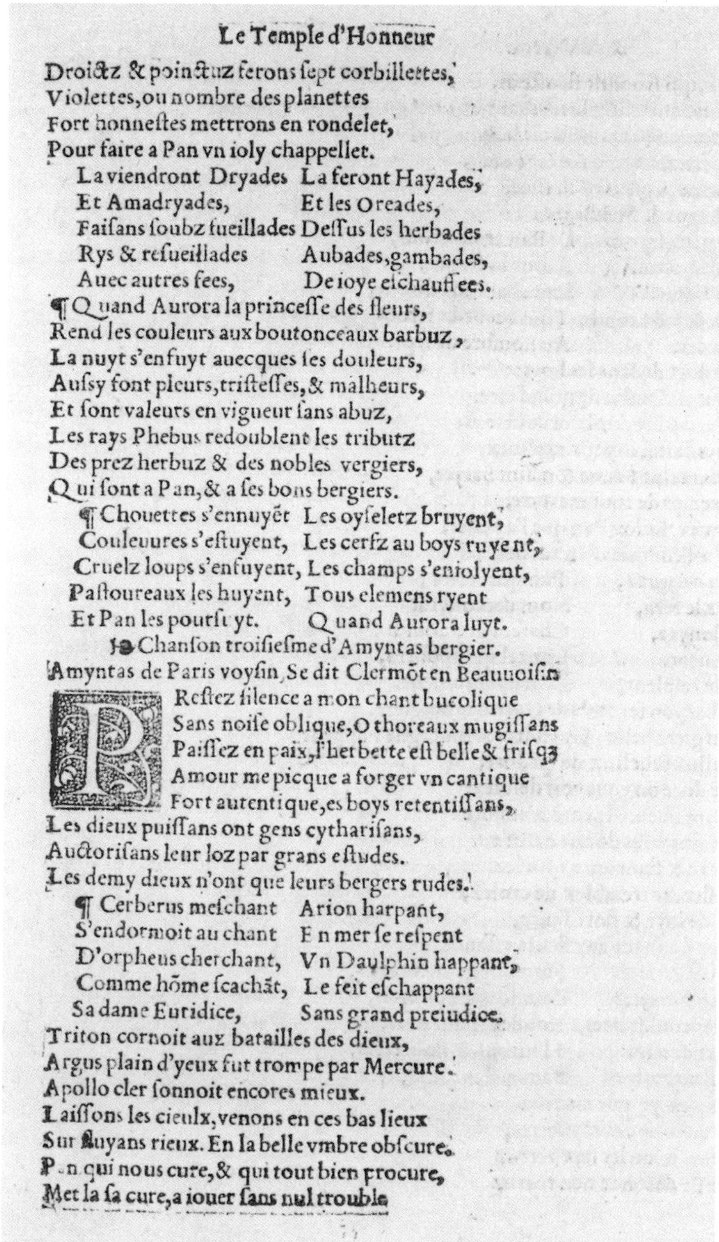

Le Temple d'Honneur

Droictz & poinctuz ferons sept corbillettes,
Violettes, ou nombre des planettes
Fort honnestes mettrons en rondelet,
Pour faire a Pan vn ioly chappellet.

La viendront Dryades   La feront Hayades,
Et Amadryades,   Et les Oreades,
Faisans soubz fueillades   Dessus les herbades
Rys & resueillades   Aubades, gambades,
Auec autres fees,   De ioye eschauffees.

¶ Quand Aurora la princesse des fleurs,
Rend les couleurs aux boutonceaux barbuz,
La nuyt s'enfuyt auecques ses douleurs,
Aussy sont pleurs, tristesses, & malheurs,
Et sont valeurs en vigueur sans abuz,
Les rays Phebus redoublent les tributz
Des prez herbuz & des nobles vergiers,
Qui sont a Pan, & a ses bons bergiers.

¶ Chouettes s'ennuyēt   Les oyselетz bruyent,
Couleuures s'estuyent,   Les cerfz au boys ruyent,
Cruelz loups s'enfuyent,   Les champs s'eniolyent,
Pastoureaux les huyent,   Tous elemens ryent
Et Pan les poursuyt.   Quand Aurora luyt.

Chanson troisiesme d'Amyntas bergier.

Amyntas de Paris voysin, Se dit Clermōt en Beauuoisin

Restez silence a mon chant bucolique
Sans noise oblique, O thoreaux mugissans
Paissez en paix, l'herbette est belle & frisq3
Amour me picque a forger vn cantique
Fort autentique, es boys retentissans,
Les dieux puissans ont gens cytharisans,
Auctorisans leur loz par grans estudes.
Les demy dieux n'ont que leurs bergers rudes.

¶ Cerberus meschant   Arion harpant,
S'endormoit au chant   En mer se respent
D'orpheus cherchant,   Vn Daulphin happant,
Comme hōme scachāt,   Le feit eschappant
Sa dame Euridice,   Sans grand preiudice.

Triton cornoit aux batailles des dieux,
Argus plain d'yeux fut trompe par Mercure.
Apollo cler sonnoit encores mieux.
Laissons les cieulx, venons en ces bas lieux
Sur fluyans rieux. En la belle vmbre obscure.
Pan qui nous cure, & qui tout bien procure,
Met la sa cure, a iouer sans nul trouble

PLATE 4.1 Parallel columns for *dizains* in the Réal/Bonfons *Temple d'Honneur et de Vertus*; Bibliothèque Municipale de Lyon, Rés. 348 128, fo. fff1v (Cliché Bibliothèque Municipale de Lyon)

plausible), again underlining Lemaire's creative investment in the text.

This visual variety is especially pronounced in the Réal/Bonfons edition, where white space reinforces the contrast between the alternating stanzas. In this section, no blank lines separate the stanzas: the only white space within the area of text on each page comes between the two columns of pentasyllables. In addition, Entendement's *double ballade* is laid out in two columns, immediately followed by the poem's final two stanzas (in the *acteur*'s voice), which, in contrast, are centred on the page.[28] Such variety, reflecting the range of different verse forms, is especially significant in comparison with the poems in the *Épître du roi* collection in the same volume. These are all presented in a single column, despite their diversity of versification; the *Temple* was apparently regarded as being of more formal interest than the *Épître* collection's constituent pieces.

More significant is the editions' use of initials. Prominent woodcut initials are employed in every edition, though their size often varies, especially in the Janot/Lotrian edition, where initials seven lines in depth appear alongside two-line *lettres de tourneure*, and other initials of every intermediate size. From a technical viewpoint, the use of initials may pose problems. In the Vérard and Le Noir editions, for example, five- and six-line initials complicate the layout of verse text, pushing verses to the right of the page and forcing several of them to run over two lines. This suggests that the use of initials in these editions was regarded as important, justifying the occasional awkward layout. The importance of large initials is confirmed by the positioning of smaller initials: small initials, *lettres de tourneure*, and paraph signs in lieu of initials all generally appear near the bottom of pages, where there is no room for larger initials, or where the printer's stock of large initials has already been exhausted on the rest of the forme. Large initials, then, are used if possible; indeed, at certain points in Vérard's edition they are used even where there is insufficient space near the bottom of a page, projecting above and below the area occupied by the text they introduce.

In every edition, the initials introduce each new speaker's discourse, clearly delineating the different voices and thus

[28] Fo. ggg5^{r-v}; Lemaire, *Temple*, 98–100, vv. 1351–417.

foregrounding the *Temple*'s characteristic polyphony. The rapid succession of different voices produces an unusually high density of initials, further underlining their structural value. In the first three editions, there are forty-nine, fifty, and forty-nine different uses of initials (including guide letters, but excluding material on title-pages): this represents an average of, respectively, almost five initials for four octavo leaves; five initials for two quarto leaves; and just over one initial per quarto page.

Moreover, the frequency of these decorative letters reinforces the text's own references to writing. They recall the letters embroidered on the personified virtues' robes, thereby drawing attention to the way in which Lemaire has appropriated this technique from Molinet. More strikingly, their use in the shepherds' epitaphs displays the wide range of thematic implications which initials assume in the Temple. Because these epitaphs are only six lines long, the initials are packed closely together on the page, giving this section considerable visual appeal (Plate 4.2).

The initials' diverse effects in these epitaphs can be examined in terms of the four distinct properties which Tom Conley has identified in the printed letter in this period. A letter may function as a *relay*, part of a written discourse which appears transparently to convey information and ideas; as an *inscription*, the trace of the mechanical processes which produce the printed book; as a *picture*, with iconic or symbolic significance; or as an *enigma*, encouraging interpretative activity without necessarily yielding meaning.[29] As letters forming part of the text in each epitaph, conveying the shepherds' praise of Pierre, the initials function as relays; this would be the case whatever their appearance. However, their unusual concentration on the page also gives them the value of pictures, because their decorative value is highlighted. This is particularly clear in the Vérard edition, whose final gathering, where the epitaphs appear, is decorated not by the grotesque *initiales à cadeaux* employed in other gatherings, but by initials from other sets. Half of these are larger than the *initiales à cadeaux*, and most include images of putti, thus presenting more pictorial interest than the grotesques; while the use of such initials in this gathering alone may be for technical reasons, it indubitably

[29] T. Conley, *The Graphic Unconscious in Early Modern French Writing* (Cambridge, 1992), 5–13. This classification is especially valuable for the study of large initials, which are more visually prominent than ordinary type.

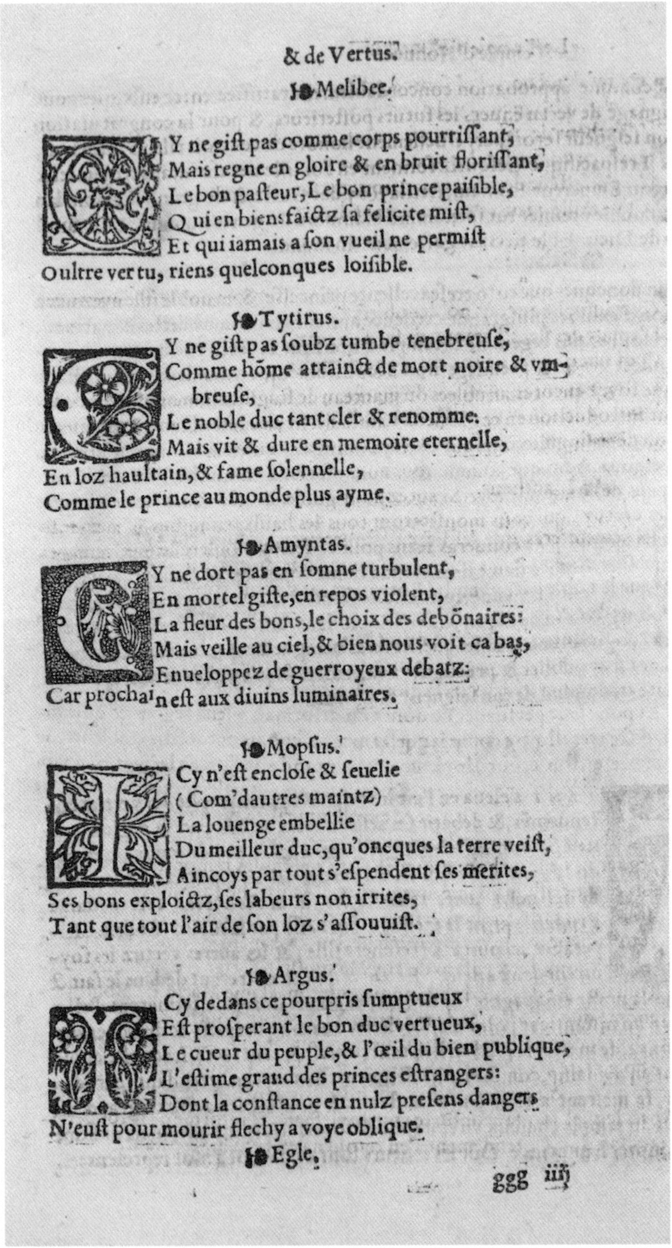

& de Vertus.

Melibee.

CY ne gist pas comme corps pourrissant,
Mais regne en gloire & en bruit florissant,
Le bon pasteur, Le bon prince paisible,
Qui en biens faictz sa felicite mist,
Et qui iamais a son vueil ne permist
Oultre vertu, riens quelconques loisible.

Tytirus.

CY ne gist pas soubz tumbe tenebreuse,
Comme hõme attainct de mort noire & vm-
breuse,
Le noble duc tant cler & renomme:
Mais vit & dure en memoire eternelle,
En loz haultain, & fame solennelle,
Comme le prince au monde plus ayme.

Amyntas.

CY ne dort pas en somne turbulent,
En mortel giste, en repos violent,
La fleur des bons, le choix des debõnaires:
Mais veille au ciel, & bien nous voit ca bas,
Enueloppez de guerroyeux debatz:
Car prochain est aux diuins luminaires.

Mopsus.

ICy n'est enclose & seuelie
(Com'dautres maintz)
La louenge embellie
Du meilleur duc, qu'oncques la terre veist,
Aincoys par tout s'espendent ses merites,
Ses bons exploictz, ses labeurs non irrites,
Tant que tout l'air de son loz s'assouuist.

Argus.

ICy dedans ce pourpris sumptueux
Est prosperant le bon duc vertueux,
Le cueur du peuple, & l'œil du bien publique,
L'estime grand des princes estrangers:
Dont la constance en nulz presens dangers
N'eust pour mourir flechy a voye oblique.

Egle,

ggg iiij

Plate 4.2 Concentration of initials in the Réal/Bonfons *Temple*; Bibliothèque Municipale de Lyon, Rés. 348 128, fo. ggg4[r] (Cliché Bibliothèque Municipale de Lyon)

complements the visual interest of the epitaphs, four of which begin with initials bearing putti. Further significance arises at the points where initials are not used. The final epitaph in each Le Noir edition begins not with an initial, but with a guide letter. This is presumably because the printer has now exhausted his stock of *I*s to mark the word 'Icy' which begins each epitaph; the only way to preserve the possibility of introducing every epitaph with an initial is to leave a space, and hope that readers will fill it. At this point, the initials assume the function of inscriptions: they highlight the material exigencies of producing a printed book.

In accentuating the book's materiality in these ways, the initials remind us that it constitutes a physical place, to which text has been added. This has important implications for Lemaire's technique of *mise en abyme*. The book as object is analogous to both Lemaire's poem and the Temple, in different ways. Like the poem, the book can be read; like the Temple, the book is written upon. Consequently, the book constitutes a kind of middle term between text and Temple. By resembling each of them, it reinforces the text–Temple equivalence, thereby overdetermining the *mise en abyme* in the poem.

The editions thus foreground various self-conscious aspects of the *Temple*: its versification; its use of an allegory indebted to, yet surpassing, Molinet's work; the motif of *mise en abyme*; and Lemaire's self-promotional use of dedications. Further effects are achieved by particular presentational features of the final two editions. Dealing first with the Réal/Bonfons *Temple*, its publication as part of a collective volume sets it in a nationalistic context. As discussed more fully below, the propagandist material in the *Traicté* and *Épître* collections, supporting Louis XII's Italian policies, simplifies the *Illustrations*' political stance and foregrounds its more narrowly Francophile aspect. Moreover, the *Temple* is immediately followed by verses by François Barat (*Œ*, iv. 438–9), which intensify the volume's pro-French flavour: they refer not to the *Temple*, but to the *Illustrations*, whose account of France's origins they praise. The political orientation of the surrounding material inevitably affects the way in which the *Temple* is read. In particular, it highlights Lemaire's praise of the House of Bourbon, which was linked to the French royal house via Anne of France: as Entendement mentions (85, ll. 990–4), Charles VII and VIII were Anne's paternal grandfather and brother.

The Janot/Lotrian edition is especially interesting, as it is the only illustrated *Temple* edition. The woodcuts used are from existing stock, often repeated, and not always especially relevant to Lemaire's text. A cut on the title-page depicts Aeneas before Apollo's temple, loosely evoking Lemaire's own temple motif.[30] Most other illustrations are likewise from an edition of Virgil: they portray pastoral subjects, including shepherds from the *Eclogues* (e.g. Plate 4.3), not all of whom appear in the *Temple*.[31] References in the text to personified virtues are often accompanied by cuts, or copies of cuts, from an edition of Gringore's *Folles Entreprises*; these depict named personifications, of little relevance to Lemaire's poem.[32] Two more figures are the two halves of a cut used in Alain Lotrian's recent edition of Villon's *Testament*.[33] Like many reused images, these cuts primarily decorate the text, and mark its divisions, rather than genuinely illustrating it. Significantly, a cut marks almost every change of speaker, underlining the *Temple*'s polyphony.

The allegorical virtues' visual potential is highlighted less than the text's pastoral aspects: of the twenty woodcuts used, six depict obviously pastoral scenes. This reinforces the bucolic/mythological orientation of the title-page, which mentions that the *Temple* includes 'les chans des bõs τ vertueux bergers suppostz de **Pan dieu siluestre/pareillement des bergieres subgectes a** Aurora'. More generally, the cuts are in an obviously classical style, and use Latin rather than French names for the shepherds (see Plate 4.3); this evokes classical and humanistic pastoral, rather than the didactic and political allegory of the native French *bergerie* tradition.[34] Consequently, attention is partially diverted from Lemaire's reworking of specifically French techniques (notably Molinet's pastoral allegories), and the *Temple*'s non-pastoral aspects.

Literary fashion is probably significant in this case. The early

[30] This cut previously appeared in Virgil, *Œuvres*, Paris, Nicolas Couteau for Galliot Du Pré, 1529 (Arsenal, 4° BL 1782), fo. DD2^{v}.

[31] See Cowling, *Building*, 7.

[32] Religion's speech, for instance, is illustrated by a cut of three figures, 'deuocion', 'richesse', and 'papelardie' (fo. E1^{v}); cf. P. Gringore, *Les Folles Entreprises*, Paris, Pierre Le Dru, 23 December 1505 (Bodleian, Douce G 386), fo. g1^{r}.

[33] Fos. B8^{r}, G1^{v}; cf. H. W. Davies, *Catalogue of a Collection of Early French Books in the Library of C. Fairfax Murray* (London, 1910), ii. 866–7.

[34] H. Cooper, *Pastoral: Mediaeval into Renaissance* (Ipswich, 1977), 79–90, discusses the *bergerie*.

PLATE 4.3 Pastoral illustration of the Janot/Lotrian *Temple*: British Library, C. 39. b. 14, fo. F8r (by permission of the British Library)

1530s in France saw a vogue for the pastoral, especially under the influence of Marot's *Premiere Eglogue de Virgile* and *Eglogue sur le trespas de ma dame Loyse de Savoye*.[35] Though the allegorical *bergerie* did not die out, it lost the dominance it had enjoyed in literary pastoral under the *rhétoriqueurs*.[36] The *Temple*'s presentation in this edition, then, was probably motivated by commercial interests; the volume appealed to current literary tastes, besides usefully exploiting a stock of woodblocks. Other aspects of Lemaire's text, such as its formal complexity, are not highlighted to the same extent as its pastoral elements. The epitaphs' layout exemplifies this tendency: each epitaph begins on a new page and is introduced by a woodcut, but the layout is not as regular as it might be, since each epitaph's heading appears on the previous page, as does the cut of Mélibée accompanying the first epitaph (Plate 4.3). Though the images reinforce the

[35] C. Marot, *Œuvres poétiques complètes*, ed. G. Defaux (Paris, 1990–3), i. 21–6, 224–31, 413–16, 625–36.

[36] Hulubei, *Églogue*, chs. 7–8.

inscriptions' visual appeal, the depiction of each new speaker underlines the text's polyphony rather than the inscriptions' materiality. Indeed, the woodcuts' presence detracts from the initials' own decorative value, further reducing the visual thematization of inscription.

Yet, despite the more particular priorities evident in the final two editions, all *Temple* editions underline a relatively wide range of the text's self-conscious techniques, testifying to its sophistication. Other works by Lemaire do not always elicit the same degree of sensitivity in presentation, as will become clear; however, the next group of Lemaire editions is of considerable interest.

Netherlandish editions, 1507–1509

Between 1507 and 1509, Lemaire was generally based in the Burgundian Netherlands, serving Margaret of Austria. The pieces he composed in this period, unsurprisingly, are his most strongly pro-Burgundian works;[37] however, they make relatively little impact in print. The *Concorde du genre humain*, *Chansons de Namur*, *Pompe funeralle*, and *Épitaphe de Molinet* (the latter two in the same volume) were each printed only once, in either Antwerp or Brussels.[38] It is uncertain to what extent Lemaire was involved in these editions' publication.

The paucity of editions of this material may be partly due to its political orientation. Whereas the *Illustrations* represented both French and Burgundian constituencies, no amount of tactful reclothing could make the pieces composed in the Burgundian Netherlands attractive to a French market. Although only the *Chansons de Namur* are explicitly anti-French, all the texts praise Margaret lavishly, in a fashion hardly appropriate in France, whose queen (and Lemaire's eventual patroness) had married Charles VIII despite his engagement to Margaret. While French and Burgundian attitudes were not necessarily mutually exclusive,

[37] See Jodogne, *Jean Lemaire*, 269–319.

[38] For these texts, see respectively J. Lemaire de Belges, *La Concorde du genre humain*, ed. P. Jodogne (Brussels, 1964); *Œ*, iv. 293–306, 243–66, 319–20. Further references to the *Concorde* will be incorporated into the text. A Dutch translation of the *Pompe* was also printed: see W. Nijhoff and M. E. Kronenberg, *Nederlandsche Bibliographie van 1500 tot 1540* (The Hague, 1919–61), no. 3377.

Lemaire perhaps prudently decided not to publish these works after entering Anne of Brittany's service. However, the few editions of these texts do not simply foreground political concerns or praise of Margaret; rather, their physical presentation highlights an intriguing variety of self-conscious techniques. My detailed discussion of the *Chansons*, *Pompe*, and *Concorde* editions is perhaps disproportionate to their relative insignificance in quantitative terms, but enables the questions they raise to be properly addressed.

The *Chansons de Namur* commemorates the victory of Ardennes peasants over French troops at Saint-Hubert in October 1507; Lemaire criticizes France, and exhorts the inert Burgundian aristocracy to resist French aggression.[39] The title specifies the text's author and his patroness, and the event concerned:

les chansons de Namur. Pour la victoire eue contre les francois A saint Hubert dardenne. ¢ Composees par Iehan lemaire de belges Indiciaire et historiographe de la tresillustre maison daustriche castille bourgoigne et de Namur. τ c̃. A lhõneur du pays Et de treshaulte τ tresclere prĩcesse madame Marguerite daustriche τ de bourgoigne. duchesse douagiere de sauoie Contesse de villars. dame de bresse τ c̃. Regẽte et gouuernante des pays de pardeca. pour le Roy son pere Cesar auguste. Tuteur et manbourg de monseigneur larchiduc Charles. Prince des espaignes. τ c̃. duc de bourgoigne de brabant. τ c̃. Conte de flandres τ c̃ Palatin de haynau. τ c̃.[40]

By mentioning Lemaire's post in connection with Margaret, the title again draws attention to the real-life social structures within which Lemaire works, structures whose concreteness is enhanced by the obtrusive realia of Margaret's and Charles's titles. Similarly, the colophon reinforces the integration of the *Chansons* with the historical event. The text is dated 'octobre mil cinq cens et sept' (fo. a6^{v}), perhaps reflecting not the actual publication date, but a wish to harmonize the date of printing with that of the battle.[41] Hence both text and poet are anchored in the real, suggesting Lemaire's prestige as court poet, and his consequent authority to recount events.

A Latin quatrain following the colophon, however, displays a different mode of literary self-consciousness:

[39] See Jodogne, *Jean Lemaire*, 299–311.

[40] Antwerp, Henri Heckert, October 1507 (BNF, Rés. Ye 942 (facsimile)), fo. a1^{r}.

[41] Jodogne, *Jean Lemaire*, 300.

Non minor bello hic labor est
Que singula multi
Fortiter egerunt
Ego prosequor oĩa solus. (fo. a6^{v}; *Œ*, iv. 307)

While the quatrain's authorship is uncertain, it expresses a notion of poetic activity consistent with that which Lemaire evinces elsewhere. It defends the value of Lemaire's writerly activity: the poem, it is claimed, deserves equal status with those whose deeds it commemorates. Ultimately, it is implied, the *Chansons* itself merits commemoration, as a difficult and therefore praiseworthy enterprise. The other editions published in this period similarly thematize Lemaire's writing, in increasingly striking ways.

The *Pompe funeralle* recounts the funeral ceremony of Philip the Handsome, Margaret of Austria's brother, at Malines in July 1507.[42] Essentially a piece of reportage, it presents little thematic or formal interest, apart from a remarkable use of the second person throughout to address Margaret: 'Et ainsi se cõmencerent vigilles. Vo⁹ madame estant presente secretemẽt en vr̃e oratoire. priant dieu deuotement. pour lame de vr̃e treschier frere vnicque. habillee en simple estat. dune faille. auec quelque nombre de voz femmes.'[43] This direct address is absent from the *Pompe*'s original manuscript version, though not from another manuscript copy.[44] While it would be expected in private correspondence, or in a manuscript with a single owner/reader, its use in a text which printing destines for a wide public helps advertise Lemaire's privileged relationship with Margaret. Moreover, the second-person form of address accords Lemaire a greater degree of presence in the text: unlike third-person narrative, it intrinsically implies the presence of a speaker.[45]

More significant than the *Pompe*, however, are the pieces printed after it in Vorsterman's edition. These are a poem entitled *De la nouvelle alliance d'Angleterre* (*Œ*, iv. 267–8) introduced by a prose account of the treaty between Maximilian and Henry VII of England which it celebrates; the *Épitaphe de Molinet*; and a letter to

[42] Jodogne, *Jean Lemaire*, 296–9.

[43] *La Pompe funeralle des obseques du feu roy dom Phelippes*, n.d.o.p. [Antwerp, Willem Vorsterman, from 15 February 1508] (BNF, Rothschild 2432), fo. B2^{r}; *Œ*, iv. 254. This edition's *terminus a quo* is provided by the date of Lemaire's letter to Charles Le Clerc (see below).

[44] BNF, Coll. Dupuy, ms. 503, fos. 144^{v}–152^{r}, and ms. fr. 22326, pp. 51–72, respectively.

[45] See É. Benveniste, 'Structure des relations de personne dans le verbe', 4–6.

Charles Le Clerc, Margaret's *trésorier des guerres*, referring to the *Épitaphe* (*Œ*, iv. 321–3). The poems are both mentioned on the title-page, which again sets Lemaire in the lineage of Chastelain and Molinet: '¢ Vng chant nouuel touchant laliance dangleterre .:. ¢ Lepitaphe de feux messire george chastelain et maistre Iehan molinet Iadis Indiciaires et historiographes. de la tres Illustre maison de Bourgoigne. Par Iehan Lemaire Belgijen leur hũble successeur disciple et loingtain Imitateur'.

Each poem's layout foregrounds its versification. *Virgulae* or *punctūs* at the hemistich highlight the *rimes batelées* in the *Nouvelle Alliance*, while paraphs mark the even lines of the *Épitaphe*, which consist of replies to questions asked in the odd lines. Lemaire's technique here is also underlined by the poem's title, which mentions that 'Ces metres alexandrins sont Interrogatifz. et Responsifz. alternatiuement' (fo. C4[v]). The attention thereby drawn to Lemaire's formal craftsmanship complements a different manifestation of literary self-consciousness in the epitaph itself, which refers to the dead poets' renown:

> Ou sont leurs monumentz et precieux tũbeaux
> ¢ En la bouche des bons. et en leurs escriptz beaux
> O dieu Cõbien vault mieulx tel tumbe q̃ de cuiure
> ¢ Dautãt que plume volle ou metal ne peut suiure. (fo. C5[r]; *Œ*, iv. 320)

Lemaire also defends the value of poetry in the subsequent letter, where he dedicates the *Épitaphe* to Le Clerc, who had offered to engrave or paint it beside Chastelain's and Molinet's tombs. Besides mentioning Le Clerc's respect for the 'princes orateurs et monarques poetes de nostre langue gallicane' (*Œ*, iv. 321), Lemaire recalls that he has succeeded Chastelain and Molinet as Burgundian *indiciaire*. He proceeds to justify his tenure of this position in a remarkable passage:

> Sil y a doncques quelque vng qui se deuille et sente offensé de ce quil na esté promeu et preferé audit office, attendu quil se cognoissoit à ce plus ydoine il mest aduis que pour mieulx persuader sa souffisance et ydoneité (non encoires assez congneue) il fauldroit quil fist vne chose. Cestassavoir prompte foy dautant de ses estudes et veilles employees, à lhonneur et seruice, de ceste tres Illustre maison comme iay fait iusques à ores en laaige ou ie suis. Non tant. Car ce seroit trop pour lui. Mais, sans plus, autant que peut monter [sc. montrer] le premier liure de la couronne margariticque des ia assez publijé et dont tu as le double. Et les deux autres premiers des singularitez de troye et de turcquie, qui ne sont encoires diuulguez . . . ce terme Indiciaire vault autant à dire comme demonstrateur. Et

est celui office si iuste quil ne requiert, faueur, ny acception de personnes, et ne se peut donner synon au mieulx monstrant la force de son engin, et sy est de tel subiection, quil ne se peut exercer par procureur. Demonstre soy doncques lautre, car on voit clerement que ie le faiz ainsi. (*Œ*, iv. 321–2)

Not only does Lemaire defend the dignity of his office, but his criticism of an unknown rival further stresses his own merits, by drawing attention to his erudition and productivity. By having the letter printed and thereby brought to the attention of a wide public, Lemaire also exploits its value as publicity. Besides reminding readers of his qualities and achievements, the letter whets the public's appetite for more of his work by alluding to the *Couronne* and *Illustrations*, neither of which had yet been published. Lemaire's letter thus displays several self-conscious techniques: the use of printing to advertise his merits and arouse interest in his forthcoming works; the thematization of literature as a demanding calling; and the assertion of his superiority over competitors.[46]

The *Concorde du genre humain* celebrates the Peace of Cambrai (10 December 1508) and lauds Margaret, who played a key role in the negotiations.[47] An intricately constructed piece, it begins with a verse section comparing the Peace to Christ's birth, and Margaret to the Virgin, exploiting the coincidence of the Peace's date with that of the Feast of the Immaculate Conception (8 December). A prose passage then praises Cambrai and Matthias Lang, Maximilian's representative at the signing, and examines the significance of the number three to the events described. In prose and verse, Lemaire then recounts a dream in which the personified Concorde du Genre Humain presents Margaret to Maximilian and requests that Margaret be granted the appellation *Auguste*; this is duly accorded, and the piece ends with an account of the historical ceremony of ratification, including verses praising Margaret delivered by pastoral personifications of major European powers.

Formally, the *Concorde* is innovative; Lemaire's compositional skill is apparent to knowledgeable readers. One of the verse sections employs an unusual variant of *terza rima*, in which tercets become quatrains rhyming *aaba*; moreover, the dream does

[46] See also Cowling, *Building*, 166.

[47] See *Concorde*, esp. 9–17; Jodogne, *Jean Lemaire*, 312–19; Brown, 'Rise', 64–74; Randall, *Building*, 85–93.

not frame the entire narrative, as is usual in political allegories, but merely forms one part of the text, and is couched in the terms of *poetrie* by being attributed to the agency of Calliope (63, l. 460). These features are not foregrounded by the layout of the edition, printed in Brussels by Thomas van der Noot in January 1508/9;[48] however, other aspects of this edition's *mise en livre* highlight Lemaire's writerly activity in significant ways.

Lemaire's name is not mentioned on the title-page, which again accumulates historical details, and establishes the link between Margaret and the Virgin: 'CY cõmence vng nouueau traictie Nomme La Concorde du gendre [*sic*] humain Compose a lhonneur de la saincte conception de la gloriose vierge/ Le Iour de la quelle fut Conclue a cãbray la tresheureu- [*sic*] paix/ Moyennant la prudence et felicite/ de madame marguerite'. But Lemaire is named immediately before the colophon: 'Jo. Marianus, apis belga, subdyaconus, fecit' (80, l. 907). His name also appears above a liminary epistle to Mercurino di Gattinara, one of the diplomats involved in securing the Peace. By underlining Gattinara's importance in the negotiations, the epistle again sets Lemaire in a network of real-life relationships, showing him to advantage as one who can address such luminaries directly.

The *prosimetrum* colophon displays a more unusual technique of self-assertion:

> Imprimé dedens Bruxelle
> par moy, Thomas de la Noot,
> le mois que le prince volt
> premier entrer en icelle,
>
> qui fut en janvier mil cincq cens et wyt, et Madame nasquit en ladicte bonne ville, le semblable mois, l'an mil CCCC quatre vingtz.
>
> Un certain imprimeur nous a fait 'Chambray' lire
> là où nous pronunchons par droicture 'Cambray':
> ou il cuide enrichir no langue, à dire vray,
> ou de sa cocquardise il nous veult faire rire. (80, ll. 909–19)

Here, poetry and publication form part of the same literary project. Van der Noot engages in a publishing as well as literary rivalry, using the colophon's prominent position to convey a

48 BNF, Rés. Ye 1043.

polemic; his criticism is directed at Philipot Le Cocq of Cambrai, who had recently published Nicaise Ladam's *La Paix faicte a Chambray*, also dealing with the Peace.[49] The inseparability of literature and printing is underlined by the use of verse, conventionally restricted to literary texts, to state the place and date of publication. Though Lemaire did not necessarily influence the colophon, it testifies to a competitive literary and commercial environment which surely contributed to the self-assertive qualities so prominent in Lemaire's printed works.

The edition's woodcuts highlight another of Lemaire's techniques. The recto and verso of the title-page both bear a cut of the Virgin, holding the infant Christ and standing on a crescent moon, being crowned with an arched (i.e. imperial) crown by two angels. The verso adds a scroll below, with the legend 'Mediatrix nr̃a. In throno dei'. These illustrations clearly reinforce the poem's opening metaphor, which presents the Peace and Margaret in terms of the Nativity and the Virgin. Indeed, the link between Margaret and Mary is reinforced by the representation of Mary with the imperial crown, currently worn by Margaret's father Maximilian.

However, the woodcut also maintains the difference between the two women. This is especially clear on the title-page verso, where the legend on the scroll can apply only to Mary. The two cannot be wholly identified with each other: despite their similarities, they never merge. In this respect, the woodcut draws attention to the way in which the text carefully distinguishes between the women, never allowing similarity to become identity.[50] The differences are apparent from the beginning, as the biblical allegory is 'solved' as soon as it is presented:

> Dieu a transmis des arcanges ung prince
> en Nazareth, qu'on dit fleur, germe ou monde,
> (c'est à Cambray, en belgicque province),
> vers une dame ayant vertu non mince,
> se vierge non, vesve au moins pure et monde. (52, vv. 87–91)

Similar distinctions are made elsewhere. Addressing the Virgin, the poet asserts that Margaret is 'ton ymaige et qui mieulx te ressemble' (55, v. 212), expressing similarity rather than identity, and indeed confirming Margaret's subordinate position:

[49] *Concorde*, 112–13 n. 201.

[50] See Randall, *Building*, 86–8.

> c'est, aprés toy, nostre plus seure planche;
> brief, ce nous est une autre Nostre Dame. (56, vv. 220–1)

By drawing attention to this irreducible difference between the women, the woodcut stresses that the historical events and their metaphorical expression are not wholly congruent. Consequently, it foregrounds Lemaire's skilful activity in weaving these disparate phenomena into a single discourse.

The infant Christ in the cut is also significant for an appreciation of Lemaire's metaphors. A historical referent for the infant cannot be identified with certainty. The Nativity metaphor implies that he represents the peace treaty, yet this equivalence only highlights Lemaire's problematic use of the term 'conception'. The Immaculate Conception to whose date Lemaire alludes is, properly speaking, the conception of Mary in God's mind and then in St Anna, not the conception of Christ in Mary.[51] Yet the poem often suggests that the Conception has been taken to mean the latter, notably in an allusion to the Peace's conception in Margaret and its subsequent Nativity-style birth: 'conceu par elle et par elle enfanté' (55, v. 208). Lemaire has exploited the ambivalence of 'conception', to establish two links: between the dates of the Peace and the Immaculate Conception, and between the Peace's signing and the Nativity. The woodcut thus draws attention to the way in which Lemaire has played upon lexical ambiguity to maximize his account's metaphorical intensity. Indeed, a profusion of overlapping metaphors characterizes the *Concorde* throughout: Margaret is later termed 'nostre clere estoille orientalle', and compared to Pallas Athena (62, l. 424; 64, v. 482). Lemaire seems to be engaging in a virtuoso performance, by accumulating as many different metaphors as possible for Margaret. By interacting with these metaphors, the woodcut sensitizes readers to Lemaire's language. It thereby reinforces the text's frequent thematization of linguistic propriety: La Concorde explains at length why the epithet *auguste* befits Margaret, while Lemaire embarks on a historical and etymological excursus to show why Cambrai is a fitting site for a reconciliation. These reflections upon language's appropriateness lend the *Concorde* a highly self-conscious flavour.

This self-consciousness is also displayed in explicit references to writing. As it progresses, the *Concorde* increasingly focuses on

[51] Réau, *Iconographie*, II. ii. 75–6.

Lemaire as literary craftsman. Cynthia Brown has shown that several passages in the *Concorde* call attention to Lemaire's writerly practice, and comment upon the text's composition.[52] An example is the evocation of creative failure preceding the praise of Cambrai: 'Cognoissant doncques que le taire m'eust esté reproché et le peu dire me causera vergoigne, ne sçay de quel part je me tourne' (59, ll. 341–3). The first-person voice here is not a witness-narrator, as often in *rhétoriqueur* pieces, but a self-conscious writer-as-craftsman. As Brown notes, these metadiscursive passages unify the text, and eventually become its central preoccupation.

Headings further underline Lemaire's presence. They emphasize the poet and his activity from the beginning, in the head-title: 'Commence la Concorde du Gendre humain, où premierement l'acteur methaphoricquement descript l'annunciacion, concepcion et nativité du fruict de paix' (51).[53] The headings even draw attention to themselves, by making readers uncertain as to whether they are in Lemaire's own voice (e.g. 'Renovation de nostre methaphore', 62) or someone else's (e.g. 'Icy descript l'acteur une vision', 63). Consequently, the headings testify to Lemaire's self-presentation, not as a narrator within the text, but as a figure who 'consciously manipulates the literary text from without',[54] and as such is susceptible to being identified with the historical author. Like the woodcut, the headings reflect the text's own tendency to thematize Lemaire's techniques, and to focus on his creative investment in language and imagery.

The *Concorde*'s closing section, where Lemaire addresses his work, not only thematizes the text's publication, but also underlines the way in which Lemaire often adopts a metadiscursive position: 'Escoute ung peu, le mien petit, mais non desagreable labeur, puisque tu ne te peuz plus contenir ou destroit de mon estude, ains te veulx eslargir par mille volumes et exposer au dangier de tous abayeurs malivoles pour cuider complaire aux vertueux qui nous favorisent, à tout le moins croy mon conseil' (79, ll. 878–83). This passage highlights the effects of printing, which, permitting texts to escape the relatively restricted circuit of manuscript transmission, exposes them to an unfamiliar public which may misunderstand them. The *Concorde*'s frequent discussions of Lemaire's

[52] Brown, 'Rise', esp. 64–71.

[53] Ibid. 64–5, discusses the headings.

[54] Ibid. 65.

practice, which draw attention to his control over the text, seem to constitute a response to this development, reducing the risk of misreading.

The Netherlandish editions, then, foreground various self-conscious techniques, though explicit reflection upon writing, and the use of print to show Lemaire's relationships with patrons and other luminaries to advantage, become increasingly important. The *Concorde* conflates the voices of narrator and historical author, highlights Lemaire's use of language and metaphor, and thematizes his writerly activity. The *Chansons*' paratext valorizes the writing of poetry, while the *Pompe* volume stresses Lemaire's calling as poet, setting him in the lineage of Burgundian *indiciaires* and underlining his office's dignity.

Editions of the Épître du roi

The tendencies exhibited in the Netherlandish editions are also manifest in the material published alongside the *Illustrations*. Of the various texts usually published in this way, I shall focus on the *Épître du roi* collection. This is a representative case, in that its presentational features and their development broadly resemble those of the other pieces accompanying the *Illustrations* (similarities with which are briefly adduced where appropriate).[55] Moreover, it contains primarily poetic texts, unlike the *Traicté* and *Légende* anthologies, and permits the presentation of the *Épître* and *Concorde des deux langages* to be compared with their presentation in manuscript, as discussed in the previous chapter.

The collection contains, in order of appearance: the *Épître* itself, the only piece explicitly mentioned on any edition's title-page; the *Couplets de la valitude de la Reine* (*Œ*, iii. 87–97), which commemorate Anne of Brittany's recovery from a serious illness;[56] Lemaire's translation into alexandrines of a Latin epitaph for Gaston de Foix, leader of French forces in Italy, killed at Ravenna in April 1512 (*Œ*, iii. 196); the *Concorde des deux langages*; Guillaume Cretin's *Plainte sur le trespas du saige et vertueux chevalier, feu de bonne memoire Guillaume de Byssipat*, lamenting Guillaume II de Byssipat, killed at the

[55] For fuller details on the other texts, see Armstrong, '*Grand Rhétoriqueur* Poetics', 263–81.

[56] See Jodogne, *Jean Lemaire*, 440–2.

siege of Bologna in early 1512;[57] and an epistle to François Le Rouge, Anne's *maistre des requestes* (*Œ*, iii. 197–8). Seven editions appeared as separate volumes between 1513 and 1524; subsequent editions generally form part of single-volume collective editions of the *Illustrations* and other material.[58]

Lemaire writes to Le Rouge that 'En la fin de mon troisieme liure des Illustrations de France, iay bien voulu à la requeste et persuasion daucuns mes bons amis, adiouster les oeuures dessusecrites' (*Œ*, iii. 197), underlining that the collection's publication with the *Illustrations* is a deliberate project. Indeed, the earliest editions' colophons indicate the extent of Lemaire's involvement in publication: 'Imprime a Paris . . . Par le commandemẽt de Maistre Ian le Maire . . . Pour Geoffroy de Marnef . . . Auquel ledict Maistre Ian le Maire/ a communicque son priuilege Royal'.[59] Such active participation in the printing of his work, already observed in Vérard's edition of the *Temple*, is typical of Lemaire. The Le Rouge epistle also evinces another characteristic trait of Lemaire's practice, his self-promotional use of letters to historical luminaries as liminary pieces in editions; this technique is employed in the *Légende*, *Traicté*, and *Épîtres de l'Amant vert*, as well as the *Temple* and *Concorde du genre humain*.[60]

The inclusion of the *Épître*, *Légende*, and *Traicté* anthologies alongside the *Illustrations* has political implications for the reading of both the volume as a whole and the *Épître* collection. Against the background of French military involvement in Italy, the anti-

[57] Cretin, *Œuvres*, 73–93.

[58] For editions after 1524, see Abélard, '*Les Illustrations*', 133–57. The earlier editions, in chronological order, are: Paris, for Geoffroy de Marnef, August 1513 (two editions: representative copies, in chronological order, are Lyons, BM, Rés. 104 686 and 317 051); Paris, [Nicolas Higman] for Geoffroy de Marnef, July 1516 (two editions, the latter probably wrongly dated: representative copies, in chronological order, are BL, G. 10249 (6) and Oxford, Codrington Library, i i 5. 7); Paris, for Jean and Enguilbert de Marnef and Pierre Viart, August 1519 (BL, 634. k. 6); Paris, for Jean and Enguilbert de Marnef and Pierre Viart, August 1521 (BNF, Rés. La^2 5 (5)); Paris, for François Regnault and Jean Petit, 1523 (BNF, Rés. La^2 6 (4)). For details of printers and dating, see Abélard, '*Les Illustrations*', 57–132, and Armstrong, '*Grand Rhétoriqueur* Poetics', 460–2. The Le Rouge epistle is absent from two single-volume editions: Paris, Philippe Le Noir, July 1524 (BNF, Rés. La^2 8), and Paris, François Regnault [in association with Philippe Le Noir], n.d. [1533–4; for dating, see Lemaire, *Traicté*, 67–8 n. 145] (BNF, Rés. La^2 7).

[59] First edition, fo. E8^r (this leaf absent from Lyons copy: details from Montpellier, BM, L391 (4) RES). Further references will be incorporated into the text.

[60] On the *Légende*'s liminary material and its development, see Brown, *Poets, Patrons, and Printers*, 53–4. The *Traicté* collection begins with a dedication to Louis XII; the *Épîtres* are introduced by a letter to the artist Jean Perréal.

Italian propaganda of the *Légende* and *Traicté* accentuates the *Illustrations*' attacks on Italian pretentions to cultural precedence in Europe.[61] Consequently, the *Illustrations*' specifically pro-French aspects are highlighted, though not necessarily to the exclusion of Lemaire's Burgundian loyalties: witness the inclusion of the *Épîtres de l'Amant vert*, concerned primarily with Margaret. As for the *Épître* collection, both its combination of texts and its publication in this composite volume foreground the pieces' political significance. The Foix epitaph and Cretin's *Plainte* commemorate French leaders killed in Italy, clearly reflecting the anti-Italian vein elsewhere in the volume. In this light, two features of the *Épître* stand out: its criticism of Julius (a theme, however, perhaps more prominent in the context of the St Petersburg manuscript, dominated by specifically anti-Julius material); and its account of France's Trojan ancestry, complementing that provided in the *Illustrations*. Similarly, the *Concorde*'s most noteworthy aspect becomes its valorization of France's culture and poetic tradition above those of Italy.[62] The *Épître* collection thus presents itself as primarily an anthology of topical, occasional pieces, promoting French military, political, and cultural concerns. Inevitably in such a propagandist context, the self-conscious features of the *Épître* and *Concorde* are attenuated, in contrast to their prominence in manuscript versions.

Besides the collection's political slant, Lemaire's exploitation of Cretin's name is also noteworthy. In the *Concorde*'s prologue, a reference to French authors is followed by the comment 'desquelz maistre Guillaume Cretin est le prince', absent from manuscript versions (4, l. 31 v. l.). Similarly, Lemaire claims in the Le Rouge epistle that he has included Cretin's *Plainte* to show 'comment la langue Gallicane est enrichie et exaltee, par les oeuures de monsieur le tresorier du boys de Vincennes, maistre Guillaume Cretin' (*Œ*, iii. 197). Lemaire is probably citing Cretin, whose French credentials were unimpeachable, to help secure his as yet uncertain

[61] See Stephens, *Giants*, ch. 4. R. Cooper, *Litterae in tempore belli: Études sur les relations littéraires italo-françaises pendant les guerres d'Italie* (Geneva, 1997), provides a broader discussion of these themes in contemporary French culture.

[62] Tuscany is exempted from the critiques of Italy: the *Concorde* ends with Labeur Historien promulgating a political union of France and Tuscany, while the Le Rouge epistle claims that 'le langage Latin, Toscan, et François, se rapportent lun à lautre, tout ainsi comme vne petite trinité' (*Œ*, iii. 197). Julius, Rome, and Venice are Lemaire's real targets.

position at Anne's court.[63] However, Cretin's poem also praises Lemaire, who is mentioned among the poets whom the narrator asks to compose laments for Byssipat:

> Abbé d'Auton, et maistre Jehan Le Maire,
> Qui en nostre art estes des plus expers.[64]

Once again, Lemaire incorporates an endorsement of his work into an edition, taking advantage of the opportunities for publicity which printing provides.

While the *Épître*'s title-pages do not specify Lemaire's authorship, those of the early editions bound into composite volumes bear a pictorial attribution, in the form of a woodcut of Lemaire's arms (Plate 4.4). These arms, which also appear in the *Illustrations* and other associated texts, present Lemaire's writerly activity in suggestive fashion. Some of his other works are alluded to: the parrot evokes the *Épîtres de l'Amant vert*, while the bees recall Lemaire's designation as 'apis belga' at the end of the *Concorde du genre humain*, which also includes a version of the legend 'Si non utile est quod facimus, stulta est gloria' (80, ll. 907–8). Moreover, the bees are symbols of literary industry;[65] in conjunction with the legend, they underline the scholarly effort required in historical and occasional writing. The cockerel, with its punning legend 'Gallis aeternum decus', echoes the *Illustrations*' praise of 'Gaul', a flexible entity which in this context appears more narrowly French. Yet the legend may be read not only as a nationalistic exclamation, 'everlasting glory to the Gauls', but also as praise of Lemaire and his project, 'an everlasting source of glory to the Gauls'. In this latter sense, it reflects the self-promotion and claims to enduring renown, already evident in Lemaire's 'liuret sommaire', which typify editions of his work. Hence, besides their political significance, Lemaire's arms thematize his activity as poet and historian, by alluding to his earlier works and literary goals.

This same combination of political and literary concerns is

[63] Lemaire employs the same strategy in book 3 of the *Illustrations*, which reproduces a letter from Lemaire to Cretin in which Lemaire claims fidelity to France: Abélard, *'Les Illustrations'*, 114–15.

[64] Cretin, *Œuvres*, 91, vv. 541–2.

[65] This topos appears, for example, in Horace, *Carmina*, 4. 2. 27–32; Horace, *The Odes and Epodes*, ed. and tr. C. E. Bennett (Cambridge, Mass., 1927), 288.

PLATE 4.4 Lemaire's arms, from the title-page of the first *Épître* edition; Bibliothèque Municipale de Lyon, Rés. 104 686 (Cliché Bibliothèque Municipale de Lyon)

apparent in a second woodcut reproduced in the editions before 1524, depicting Anne of Brittany in the guise of Juno (Plate 4.5). Originally designed for Book 3 of the *Illustrations*,[66] the cut flatters Anne by associating her with the 'Deesse de toute richesse et opulence' (*Œ*, i. 232), and draws attention to the *Couplets*, which focus on Anne. At the top right is Mercury, god of eloquence, with a cockerel on his head, visually reflecting the way in which, in the *Épître* collection, Lemaire's talents are employed in France's service.

Some headings in the editions highlight Lemaire's attention to form and intertextual allusions. The *Épître*'s title mentions that the piece is 'responsiue a celle que Monseigneur Reuerend Prelat Labbe Dangle/ en Poictu/ [*sic*] Damp Iehan Dantõ [*sic*] Croniqueur du Roy Treschrestiẽ Loys .xii. Nagueres enuoiee [*sic*] audict seigneur/ de la part Dhector de Troye' (fo. A2^{r}). Consequently, it enables Lemaire's jokes about Hector's fluency in French, and unfamiliarity with modern warfare, to be appreciated as play upon d'Auton's text, though the absence of the latter renders this dimension much less apparent than in the St Petersburg manuscript. The Foix epitaph's title underlines Lemaire's skill as versifier and translator: 'Epytaphe de feu de clere memoire tresnoble ⁊ puissant prince monseigneur Gaston de foix. duc de Nemours/ Comte destampes. ⁊c. Translate de Latin en Francois Par Ian le Maire Secretaire et Indiciaire de la Royne/ Et rendu le Francois correspondant au nombre des syllabes du Latin' (fo. B5^{v}). However, the features highlighted in these headings are perhaps less prominent than the texts' political import, foregrounded by the patriotic nature of the volume as a whole.

Layouts vary relatively little between editions. The pre-1524 editions usually begin each text on a new page,[67] and present verse in a single column, visually contrasting with the much wider blocks of prose. To an extent, these features underline the variety of forms which the collection presents—prose, verse of different kinds, *prosimetrum*—and thus indicate the range of Lemaire's compositional talents. After 1524, however, texts rarely begin on a new

[66] See Abélard, '*Les Illustrations*', 110; R. Mortimer, *Harvard College Library Dept. of Printing and Graphic Arts Catalogue of Books and Manuscripts*, i: *French 16th Century Books* (Cambridge, Mass., 1964), ii. 442–3.

[67] Higman's two editions are less consistent in this respect: the texts are crammed onto fewer leaves than in other early editions.

PLATE 4.5 Anne of Brittany as Juno, from the first *Épître* edition; Bibliothèque Municipale de Lyon, Rés. 104 686, fo. A1[v] (Cliché Bibliothèque Municipale de Lyon)

page. In numerous editions, moreover, both verse and prose are laid out in two columns;[68] this layout renders prose sections very little wider than verse, attenuating the contrast between the two. Philippe Le Noir's 1524 edition is particularly striking in this respect, since prose in other parts of the volume is in long lines. This visual homogenization is attenuated, however, in the editions of Mareschal and du Ry: here alexandrines remain in a single column, unlike other verses.

The last two sixteenth-century editions return to a more varied layout. In the Réal/Bonfons edition of 1548–9, previously discussed in connection with the *Temple*, the Le Rouge epistle is printed in a larger typeface than the other material. This typeface appears elsewhere only in headings, colophons, and some liminary pieces, and thus distinguishes Lemaire's correspondence from his poems. Yet the *Épître*, like the *Temple* in this edition, is followed by Latin verses by François Barat, praising Lemaire's historiography in the *Illustrations* (*Œ*, iv. 437–8). Despite the typographical variation, then, the *Épître* material is primarily presented as an adjunct to the *Illustrations*. Still more visual variety characterizes the edition printed in Lyons by Jean de Tournes in 1549, which reproduces not only the *Illustrations* and other texts usually published with it (including the *Légende*, for the first time in a single-volume edition), but also Lemaire's *Couronne margaritique* and *Contes de Cupido et Atropos* (*Œ*, iii. 39–67).[69] Throughout the volume, verse is printed in italics, and prose in roman type; this use of italics, common in de Tournes's editions, highlights the anthologized texts' formal diversity.[70] However, the material which in other editions forms the *Épître*, *Légende*, and *Traicté* collections is very differently organized here. The *Légende* and *Traicté* proper are placed together, as pieces of anti-Italian propaganda, alongside the *Blason des armes des Vénitiens*, a short anti-Venetian poem

[68] Lyons, Jacques Mareschal, 1 June 1524 (Cambridge, Clare College, D'. 1. 30); Paris, Philippe Le Noir, July 1524 (BNF, Rés. La^2 8); Lyons, Antoine du Ry, 1528 (BL, C. 38. g. 12); Paris, François Regnault, 1528 (BNF, Rés. La^2 8 A); Paris, Nicolas Higman for Ambroise Girault, 1533 (Cambridge, Trinity College, Grylls 6. 236); Paris, François Regnault [in association with Philippe Le Noir], n.d. [1533–4] (BNF, Rés. La^2 7). See Abélard, '*Les Illustrations*', 133–44, 149–50.

[69] Lyons, BM, Rés. 108 167. See A. Cartier, *Bibliographie des éditions des de Tournes, imprimeurs lyonnais* (Paris, 1937–8), i. 280–4.

[70] Most verse and some of the prose is laid out in two columns; the Foix epitaph is in a single column.

appearing in most *Traicté* editions (*Traicté*, 273–4). The *Épître* and associated pieces, together with the other texts from the *Légende* and *Traicté* collections, directly follow the *Illustrations*. Most of the *Épître* material immediately follows book 3 of the *Illustrations*; the Foix epitaph and Le Rouge epistle follow the *Plainte du Désiré* and *Regretz* from the *Légende* collection. The pieces are thereby set in an even clearer context of political, historical, and occasional writing, which to a degree attenuates their self-conscious features.

In some later editions, the *Épître*'s title-page is illustrated. That of Philippe Le Noir's 1524 edition depicts a figure labelled as Hector, in a block probably first used in Michel Le Noir's *Neuf Preux* of 1507 (Plate 4.6).[71] Philippe has clearly taken the opportunity the *Épître* afforded to re-employ a stock image, which draws attention to the *Épître*'s recipient rather than to its ostensible author Louis XII, or to its historical author Lemaire. This underlines the piece's mythological aspects and references to France's Trojan roots, rather than its anti-papal polemic, or the ingenuity of Lemaire's response to d'Auton. The rest of this volume similarly focuses on things Trojan: book 1 of the *Illustrations* contains various woodcuts on antique subjects, and Latin verses on the fall of Troy.[72] Such elements signal a shift of emphasis from French history onto the story of Paris; this reflects the Trojan and mythological focus of the *Épître*'s title-page, though the image of Hector may also highlight the anti-Greek sentiments expressed in the *Épître* and *Illustrations*.

Philippe Le Noir employs a similar cut in his next edition, produced in association with François Regnault in 1533–4; this bears a copy, or a later state, of the cut of Caesar from the 1507 *Neuf Preux*.[73] It is uncertain whether the cut represents Hector or Louis.

[71] *Les Neuf Preux*, Paris, Michel Le Noir, 3 December 1507 (BL, C. 107. g. 6), fo. N6^v.

[72] Woodcuts include the Judgement of Paris (fo. B4^v), from Dares Phrigius, *De excidio Troie*, Paris, Nicolas Des Prez for Pierre Gaudoul, 1520 (BNF, Rés. J. 1719), fo. A2^r. See Abélard, '*Les Illustrations*', 134. The verses (fos. b2^v–3^r) are from Virgil, *Aeneid*, 2. 54–6, 261–5, 554–8 (*P. Vergili Maronis opera*, ed. R. A. B. Mynors (Oxford, 1969), 128, 135, 144), and Baebius Italicus, *Ilias Latina*, ed. M. Scaffai (Bologna, 1982), 273–6, with some variations. On the *Illustrations*' mythological dimension, see A. Moss, *Poetry and Fable: Studies in Mythological Narrative in Sixteenth-Century France* (Cambridge, 1984), ch. 2; Campangne, *Mythologie*, 136–66.

[73] For the history of the 1507 blocks, and a reproduction of the Caesar image, see Sotheby's (auctioneers, London), *The Collection of Otto Schäfer Part II: Parisian Books, Sold by Order of the Dr. Otto-Schäfer-Stiftung e.V., London, Tuesday, June 27, 1995* (Abingdon, 1995), no. 182. On this edition of the *Illustrations*, see Abélard, '*Les Illustrations*', 138–40.

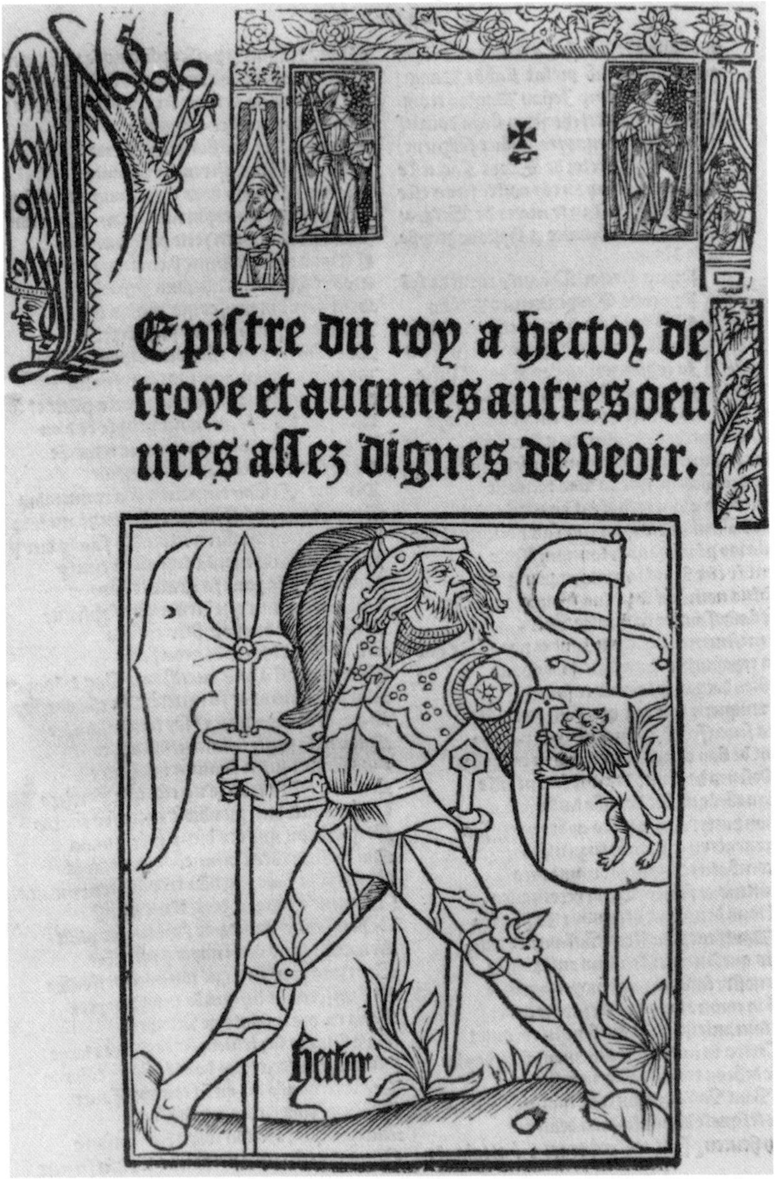
Epistre du roy a hector de troye et aucunes autres oeu ures assez dignes de veoir.

PLATE 4.6 Title-page of the *Épître* in Philippe Le Noir's 1524 *Illustrations* (Bibliothèque Municipale de Bourges)

However, the figure's shield bears the imperial double-headed eagle, which may suggest the imperialist aspirations of either Louis XII or Francis I, France's king at the time of this edition's publication. The Lyons editions of Mareschal and du Ry bear a stock cut of a mounted knight and two dogs on the verso of their title-pages.[74] Again, the image's referent is uncertain, but the cut also appears in book I of the *Illustrations*, providing a link with Trojan history which may suggest it represents Hector in this context.

The *Épître* editions, then, do partly acknowledge the texts' self-conscious aspects, although their political implications are inevitably stressed, and some later editions produce effects of their own. Lemaire's careful nuancing of his political attitudes, and consequent appeal to as wide an audience as possible, are evident in the context of the *Illustrations* and associated pieces alongside which the *Épître* is published. Within the collection, Lemaire's characteristic use of printing to further his interests and raise his profile is apparent in various respects. The Le Rouge epistle, and Cretin's *Plainte*, exemplify Lemaire's advertising of relationships with patrons and other luminaries, which helps guarantee his authority as scholar and poet; the woodcuts in early editions highlight Lemaire's writerly practice; some pieces' titles draw attention to particular compositional features. Besides promoting Lemaire's work, these features thematize and reflect upon his writing; they consequently complement the reflexive tendencies prominent in Lemaire's later work. These have been noted in the *Concorde*, within the *Épître* collection, but are equally striking elsewhere in the *Illustrations* volumes, especially in the *Légende* and *Traicté*, whose first-person narrative voice self-consciously guides readers through his political arguments and painstakingly makes clear the historical proofs underpinning them.[75] Hence the *Épître* editions illustrate the ways in which material published with the *Illustrations* continues the trend noted in the Netherlandish editions, away from formal preoccupations and towards a focus upon authority and the practice of writing.

[74] The cut copies that on the title-page of *Le Petit Artus de Bretaigne*, Lyons, Jean de la Fontaine, 4 July 1493 (ÖNB, Ink. 11. F. 35); see G. Susong, 'Les Impressions arthuriennes françaises (1488–1591) et la grande rhétorique', 190.

[75] Brown, *Shaping*, 58–63, 93–9.

Conclusions

Although generalizations about editions of Lemaire's poetry are reductive, in view of their number and variety, some broad tendencies are identifiable. The earliest editions, of the *Temple*, emphasize a wide range of self-conscious techniques, including formal intricacy, *mise en abyme*, and Lemaire's use of liminary material. In the Netherlandish editions, the thematization of writing, and the use of print to help further Lemaire's career, assume increasing importance. The *Épître* editions continue this development, reflecting the priorities stressed in editions of the *Illustrations* and associated material: Lemaire's links with guarantors of his merit; the delineation of a political stance; and explicit reflection upon aspects of Lemaire's literary and historiographical activity. After Lemaire's death, re-editions of previous publications generally imitate their predecessors, though their presentation sometimes compromises Lemaire's artistic or polemical strategies: de Tournes's *Illustrations* edition destroys the coherence of the *Épître*, *Légende*, and *Traicté* collections, while the woodcuts in some later *Épître* editions cannot always be clearly related to themes in the text. Printers inevitably take initiatives determined by their own aesthetic or commercial priorities: thus, for example, Janot and Lotrian annex the *Temple* to the literary fashion for pastoral in the early 1530s, fruitfully employing stock woodcuts in the process. In such cases, texts' self-conscious aspects are foregrounded selectively, if at all.

Indeed, Lemaire editions as a whole highlight a narrower range of texts' self-conscious features than do manuscripts. The *Épître*'s intertextual play off d'Auton, and the sophisticated allegory of the *Concorde des deux langages*, are considerably less prominent in print than in the manuscripts previously examined. As my final chapter will now discuss, these characteristics of Lemaire editions contrast with those displayed by editions of the work of Jean Bouchet, whose career differs from Lemaire's in certain crucial, interrelated respects. He was an active writer for a generation longer than Lemaire; he relied less upon patronage; his output depended more heavily upon printing; and he participated more consistently in publication.

5

BOUCHET'S POETRY

Introduction

Jean Bouchet's prolific literary career spanned some sixty years, during which print became the dominant medium of transmission for French poetry. Very few manuscripts of Bouchet's work survive, perhaps partly because Bouchet was never a court poet proper and was therefore less likely to invest in presentation manuscripts for his works.[1] However, unlike Lemaire, Bouchet was able to participate in the publication of most of his works long after their first editions. After Paris printers published pirate editions of his earliest poetry, he had the first editions of his material printed first in Paris and then, from 1518, in his home town of Poitiers, though later editions often appeared elsewhere. First editions usually bore a privilege; later in Bouchet's career, these were granted to Bouchet himself rather than to printers.[2] Some texts also contain suggestions that Bouchet's role in publication was very active. A case in point is the preface to his *Epistres morales et familieres*, a collection of verse epistles addressed to either sections of society or real-life acquaintances, published in 1545. Here, Bouchet claims that he is embarking upon a programme of publishing definitive editions of his work, to prevent unscrupulous printers from corrupting his texts: 'aduerty que par l'auarice d'aulcũs Imprimeurs on auoit peruerty voire falsifié aulcũs de mes liures . . . I'ay par le congé, octroy & permission du Roy faict imprimer de nouueau lesdictes Epistres Morales & Familieres, & aultres petiz traictez en prose & vers, & commencé a faire reimprimer au vray les aultres.'[3] Bouchet's continual involvement in publication is also shown by his tendency to rework texts,

[1] On Bouchet's patrons, see Britnell, *Jean Bouchet*, 5–11. Some presentation manuscripts, however, presumably existed; on manuscripts of Bouchet's work, see ibid. 321, 326, 330–2; A. Armstrong, 'More Manuscript Copies of Jean Bouchet's Verse: Mss. B. N. fr. 2206 and 2231'.

[2] Britnell, *Jean Bouchet*, 300.

[3] J. Bouchet, *Epistres morales et familieres du Traverseur*, ed. J. Beard (Wakefield, 1969), fo. ã6^r.

both to take account of changing conventions of versification and to reflect new historical circumstances. Thus, for example, the *Déploration de l'Église militante*, first published in 1512 as a piece of pro-Gallican propaganda, was altered in 1525 and 1545. The first revision formed part of Bouchet's moralizing anthology *Les Opuscules du Traverseur*, whereas the original *Déploration* was published separately: the text's calls for church reform are related to the recent spread of Lutheranism, and its lyric caesuras eradicated in accordance with Bouchet's current practice. Bouchet's second revision was again subsumed into a larger collection, the entire *Opuscules* being incorporated into a volume entitled *Les Genealogies, effigies et epitaphes* (discussed below), while some of the *Déploration*'s criticism of the Pope was attenuated.[4] These tendencies suggest a high degree of self-consciousness on Bouchet's part, an awareness of the importance of a correct text and proper distribution. However, this is by no means the only self-conscious feature of his work. Bouchet's historical writing, for example, often highlights his scholarly activity in finding and reconciling source materials, while the *Epistres* contain an unprecedented amount of apparently personal poetry.[5] I shall examine the ways in which the presentation of some of Bouchet's works helps shape and convey such features. The texts selected enable key aspects of Bouchet's output to be addressed: they cover a broad chronological spectrum; have different forms and functions; are published both in Poitiers, in editions presumably influenced by Bouchet, and elsewhere, presumably without his co-operation; and exhibit the reworkings characteristic of Bouchet's practice. I begin by analysing the *Regnars traversant*, from the early sixteenth century, later reworked in manuscript form; I then proceed to discuss the *Labirynth de Fortune*, first printed in 1522, and the *Anciennes et modernes genealogies des roys de France*, published several times between 1528 and 1545. Similarities and contrasts between these texts and Bouchet's other works are occasionally noted.

[4] J. Bouchet, *La Déploration de l'Église militante*, ed. J. Britnell (Geneva, 1991), 12–40; G. Bellati, 'Bouchet e l'abolizione della *coupe féminine*'.

[5] See Britnell, *Jean Bouchet*, 4–14, 118–20. LeBlanc, *Va lettre va*, 54–5, relates the content and organization of the *Epistres* to Bouchet's involvement in their publication.

Les Regnars traversant

Bouchet's early didactic work, the *Regnars traversant*, was first published in Paris, by Anthoine Vérard, in 1503–4.[6] However, the *Regnars*' printed form hardly did full justice to its author. Bouchet's name, then unknown, was replaced on the title-page by that of Sebastian Brant, presumably to help sales by exploiting the *Narrenschiff*'s contemporary popularity in France. Moreover, the *Regnars* itself constituted only the first section of Vérard's edition; over half the volume was occupied by various moralizing texts, some by Bouchet and some by other poets. Bouchet filed a lawsuit against Vérard, and the *Regnars* was also involved in a suit brought by André de La Vigne against Michel Le Noir, who had published unauthorized editions of both the *Regnars* and material by La Vigne.[7] However, later editions of the *Regnars* reproduce the early editions' text, while Bouchet's manuscript revision of the *Regnars* alone was never printed.[8]

The different *Regnars* versions thus contrast sharply with each other, in respect of textual details, their modes of transmission, and the degree to which Bouchet influenced their physical presentation. Moreover, the *Regnars* is particularly fruitful for analysis because it is illustrated by complex woodcuts, which interact in intriguing ways with the piece's governing metaphor. Consequently, the piece provides a fruitful opportunity to examine the ways in which various material forms may foreground a text's diverse self-conscious aspects.

In its initial printed form, the *Regnars* is a *prosimetrum* text in thirteen chapters. A first-person narrator vituperates against human vice and duplicity, using foxes as a metaphor for humankind.[9] The opening sections establish these themes and techniques. A

[6] Bodleian, Douce B 787. Further references, unless otherwise stated, are to this edition, and will be incorporated into the text.

[7] On the *Regnars* volume's early publication and contents, see É. Picot and A. Piaget, 'Une supercherie d'Antoine Vérard: *Les Regnars traversans* de Jehan Bouchet'; G. Bellati, 'Étude sur les *Regnars traversant* de Jean Bouchet'; Britnell, *Jean Bouchet*, 304–5; Brown, *Poets, Patrons, and Printers*, 21–33; Winn, *Anthoine Vérard*, 78–81.

[8] Later editions are: Paris, Anthoine Vérard, n.d. (BNF, Vélins 1103); Paris, Michel Le Noir, 21 May 1504 (Bodleian, 4° Buchanan d. 36); Paris, Philippe Le Noir, 1522 (BSG, 4° Y. 426^{25} inv. 589 rés.); Paris, Philippe Le Noir for Denis Janot, 25 January 1530 (Bodleian, Douce B subt. 248). See Britnell, *Jean Bouchet*, 304–6; Mortimer, *Harvard*, i. 142–4.

[9] Britnell, *Jean Bouchet*, 81–9, examines the *Regnars*.

modified extract from Brant's Latin poem *De spectaculo conflictuque vulpium Alopekiomachia* (fo. a1^{v}) opens the work by linking the fox's dual colouring, denoting its duplicity and cunning, to human deceit.[10] In chapter 1, the narrator recounts the abuses he has witnessed in society. He has also seen foxes at large, together with wolves and a lynx, with dire biblical prophecies inscribed above them:

> Sur ces figures linx/ loups et renardeaulx
> Qui me furent a les veoir fort nouueaulx
> Pour leurs habitz et facon inomye [*sic*]
> En regardant ie vy des escripteaulx. (fo. a3^{v})

These biblical texts imply the foxes have a symbolic value, as in Brant's poem. The next two chapters confirm this. In chapter 2, the narrator explains that the foxes indicate the world's parlous state: he refers to 'lesditz renars et significacion diceulx par lesquelz ie cõgnoissoye visiblemẽt le monde estre pire qui [*sic*] ne fut oncques' (fo. a3^{v}). Chapter 3 then elucidates 'la complexion du renart auec la significatiõ dicelle' (fo. a4^{v}). Foxes represent depraved humans, especially people who use deception:

> Dautres renars de ceste sorte
> y a dessoubz le firmament
> Qui mordent par parolle morte
> Soubz vmbre daymer loyaulment . . .
>
> Ceulx cy sappellent detracteurs
> Faulx rapporteurs et langagiers. (fo. a5^{r})

Foxes were traditionally associated with deceit and malice, in bestiaries, proverbs, and literary texts such as the *Roman de Renart*.[11] Presenting humans as foxes, then, implies they share these characteristics. Bouchet emphasizes the difference between the appearance and the reality of hypocrites and slanderers. Chapters 4–13 then explore the fox metaphor in more depth, expanding upon chapter 1's brief account of the narrator's vision. In each of these chapters, the narrator sees a group of foxes, accompanied by one or more biblical inscriptions in Latin; a woodcut preceding each chapter depicts the foxes and text. The narrator interprets

[10] S. Brant, *Varia carmina*, Basle, Johann Bergmann von Olpe, 1498 (Bodleian, Auct. 2 Q 6. 33), fos. h1^{v}–3^{v}.

[11] See, for example, J. Flinn, *Le 'Roman de Renart' dans la littérature française et dans les littératures étrangères au Moyen Âge* (Toronto, 1963); Hassig, *Medieval Bestiaries*, ch. 6.

each group of foxes as signifying a group of sinners, whom he addresses in a prose critique and a verse *Exhortacion* to reform.

The cuts of foxes were designed especially for the *Regnars*, and generally follow the text closely.[12] However, in view of the unauthorized publication of Bouchet's work, and in the absence of evidence regarding Vérard's relationship with Bouchet, it is uncertain whether they are based on the author's own sketches or instructions. What is certain is that the *Regnars* explicitly calls for illustration: the narrator exhorts sinners 'pensez aux figures dessus declairees et ne regardez seulement la paingture' (fo. e5[v]). *Figure* may denote either a visual image or a non-visual metaphor, but here, in conjunction with 'regardez' and 'paingture', it unambiguously refers to illustrations. Pictures, then, were part of Bouchet's project, whether or not the cuts derive from his designs.

The fox images appear between the title and text of each relevant chapter. While very common in contemporary books, this arrangement has important implications for the illustrations' role in the reading process. Each image acquires an enigmatic quality, because readers see it before the following chapter explains the scene depicted. The chapter title orientates their interpretation of the image, as it precedes the cut and states the subject of the chapter. But the title does not fully explain the image's significance; rather, readers must work out how the image can be interpreted to mean what the title suggests, on the basis of what they already know about foxes, from cultural tradition and from the *Regnars*' opening chapters. The biblical citations give further interpretative clues, and suggest connotations. Readers' provisional interpretations of the image provide them with a set of expectations with which to read the subsequent chapter; however, the chapter may supply new information, and make readers reassess the image.

Chapter 4 exemplifies this process. Its title, 'Des murmures/

[12] Vérard's volume also contains many other cuts, largely reused from stock and therefore not closely related to the texts they illustrate: J. Macfarlane, *Antoine Vérard* (London, 1900), 91, identifies several examples. The only other cut in the *Regnars* proper depicts a man writing (fo. a2[r]), a common image which Macfarlane identifies in other Vérard books (136). The fox illustrations are indebted to the original German flysheet version of Brant's *De spectaculo*, printed under the title *Von dem Fuchshatz* in 1497, with a large woodcut depicting foxes engaged in various nefarious activities: see S. Brant, *Flugblätter*, ed. P. Heitz (Strasbourg, 1915), no. 18. Britnell, *Jean Bouchet*, 82–3, discusses in detail the links between Brant's and Bouchet's pieces.

PLATE 5.1 Anthropomorphized foxes: chapter 4 of Vérard's first *Regnars traversant* edition; Bodleian Library, Douce B 787, fo. a6^{r} (reproduced by permission of the Bodleian Library, University of Oxford)

blasphemes/ et enuies du commun peuple et de la dissolucion de ses habis causans sa destruction', seems to identify the three foxes in the woodcut as the 'peuple' (fo. a6^{r}; Plate 5.1). Yet the reader must establish what the foxes are doing: are they strewing tails over the ground, or gathering them up? How does this represent discord? Do the tails denote 'murmures'? Why are there three foxes rather than, say, two or four? The accompanying citation (Lam. 5: 18) must itself be read metaphorically: it presents the foxes' misdeeds as the cause of impending doom, and provokes enquiry as to what Mount Zion represents. In the chapter, the narrator explains that the foxes are strewing tails, and interprets them as 'les laboureurs mecaniques/ marchans et bourgoys de villes et citez' (fo. a6^{v}). Mount Zion stands for the world, while the tails denote sins: 'les queues quilz sement tous les iours par lesquelles ie entens les grans pechez que ordinairement ilz cõmettent sans estre corrigez ont tellement empunaisy le monde quil en aura a souffrir et endurer' (fo. b2^{r}). Jeremiah, we are told, confirms this reading: 'le mons de syon se perist par ce que les renars ont chemine par icelluy' (fo. b2^{r}). Why there are three foxes remains unexplained: the text only mentions 'plusieurs' (fo. a6^{r}). The woodcut's interpretation may be explained by the three groups constituting the 'peuple', or by the people's three vices, envy, blasphemy, and 'dissolucion des estatz' (fo. b1[r]); but here readers are left to their own devices. However, the text adds an element absent from the woodcut, by implicitly comparing foxes' inherent ambiguity to humans' disrespect for the discreteness of estates: 'Difficile est a congnoistre lestat des hõmes a leurs habitz' (fo. b1[v]). Text and illustration contain their own overtones: the text makes readers re-evaluate the image.

The woodcuts, then, require intricate analysis. By presenting the visions as if seen through the narrator's eyes, they invite readers to engage in the same interpretative activity as the *acteur*. The important relationship between the cuts and the fox imagery is underlined by the ambivalence of *figure*: its dual sense, visual and metaphorical, encourages readers to consider how the illustrations might signify metaphorically.

The cut accompanying chapter 4 also elicits a metaphorical reading because it presents foxes walking on their hind legs like humans, anthropomorphizing them. The foxes are not behaving as they do in real life; the scene does not make literal sense. Consequently, it must be interpreted metaphorically. Furthermore, such anthropomorphic images, whether presented visually or in a verbal description, incorporate clues to their meaning. They depict foxes with human characteristics, so that in a sense they portray humans as well as foxes. Other woodcuts depict foxes behaving as foxes normally do; these contain no inherent clue that the foxes represent humans, although it is clear from the context of the *Regnars* that this is precisely what the foxes do. Both types of image, however, make similar demands upon readers: their relevance to the moral discourse of the chapter title and biblical texts must be established, a process which frequently poses delicate interpretative problems.

Chapter 6 presents a vision of foxes behaving like foxes. Its title identifies a general group of sinners and its subgroups: 'Des folz esperans q[i] sont deceuz par paresce/ par oppiniatrete/ par confiance et deffiance de fortune/ et par trop se confier es biens de nature es biẽs de grace/ es biens de fortune/ et es cours des princes' (fo. b4[v]). The woodcut depicts two foxes by a bush, snarling at several dogs surrounding them (fo. b5[r]; Plate 5.2). Behind the bush are three more foxes, with another apparently

PLATE 5.2 Foxes behaving as foxes: chapter 6 of Vérard's first *Regnars*; Douce B 787, fo. b5^{r} (reproduced by permission of the Bodleian Library, University of Oxford)

disappearing into it. It seems possible to interpret the image in terms of the title, since the foxes' hopeless resistance may aptly symbolize misplaced hope. However, the accompanying citation (Jer. 2: 19) adds a possibly discordant perspective: it corresponds to the title by suggesting the foxes' ultimate defeat, but also asserts the theme of malice, visually reinforced by the snarling foxes' prominent sharp teeth. Readers must turn to the text to establish the relevance of malice to foolish hope.

Yet the text does not wholly correspond to the woodcut. It mentions only one fox, who, chased by dogs, 'sarresta pres vng buysson/ et mouilla sa queue de sa [*sic*] date. Et quant les chiens cuyderent approucher de luy les aspergea et arrousa de sa queue/ et leur en bailla par les dens. Et incontinent au moyen de la punaisie du date se arresterẽt les chiens de courir/ mais tantost recommẽcerent. Et finablement prindrent et tuerent le renart' (fo. b5^{r}). Unable to represent the entire narrative in one scene, the woodcut designer has selected a single moment, omitting the fox's eventual death, which the citation already suggests, and the urine motif, which would be difficult to depict.

The narrator's interpretation complicates matters further: 'ce renart representoit les ioncheurs du temps qui court. Lesquelz comme lon dit cõmunemẽt passẽt la queue par le bec. Aussi nous sont icy demõstrez les folz esperans qui fondent leurs fortunes sur la glace dune nuyt' (fo. b5[r]). Nothing in the image, title, or citation has suggested that the foxes represent *joncheurs* (deceivers), apart perhaps from *malicia* in the biblical text.[13] Yet *joncheurs* appear more appropriate than 'folz esperans' to the theme of malice: the narrator explains that *joncheurs* are 'gens plains de malice et damnee cautelle' (fo. b5[r]). But the narrator later claims that the scene is more properly interpreted as representing 'folz esperans' (fo. b5[v]). He even links 'folz esperans' with the theme of malice, by comparing them to the fox who hopes to escape the dogs: 'ilz sont menez a cõfusion tout ainsi que est le regnart par sa queue de laquelle il se cuide malicieusement sauluer pour faire son date sur icelle et en arrouser les chiens qui le suyuent' (fo. c1[v]). Nevertheless, while the image appears apt for 'folz esperans', the citation seems better suited to *joncheurs*, especially because it comes nearest to being translated in that context: 'quãt les matieres sont esclaircies et dilucidees se treuuent confus et sont reprouchez de leur malice mesme' (fo. b5[r]). Different verbal and visual elements suggest competing readings of the scene. The interaction between text and image questions the narrator's interpretation, inviting readers to judge its legitimacy.

Furthermore, nowhere does the text reflect the woodcut's suggestion that some foxes (behind the bush) escape the dogs; nor does the chapter interpret the dogs or the fox's urine. The text–image discrepancies, and consequent interpretative problems for the reader, show how the text does not fully elucidate each vision; readers always have gaps to fill. Indeed, the text suggests that the metaphorical vision itself cannot account for all the sinners mentioned in the title. The narrator says of the indolent: 'ne semblent pas telles gẽs au renart q[i] fuyt et quiert tousiours moyens pour euader les perilleuses hurtes/ mais ressemblẽt a celuy qui attend la gueule bayee et les mains lyees se la viande luy entrera en la bouche' (fo. b5[v]). Another fox metaphor must be introduced to explain the original metaphor's implications; the image and citation are no longer relevant in this case. Such anomalies between text, title, and image invite active analysis by readers.

[13] *Malice* often meant 'deceit' in Middle French: R. Cotgrave, *A Dictionarie of the French and English Tongues* (London, 1611), fo. Eee2[r].

PLATE 5.3 The final vision: chapter 13 of Vérard's first *Regnars*; Douce B 787, fo. e4v (reproduced by permission of the Bodleian Library, University of Oxford)

The series of complex text–image interactions culminates in chapter 13, 'Des punicions de dieu: et de sa fureur' (fo. e4v; Plate 5.3). Its accompanying woodcut depicts a forest clearing: in the centre, facing the reader, is a lynx carrying a sword and bow; a dragon occupies the foreground; a lion and bear enter the scene from the right; and a man sleeps in the left background. The title suggests the animals represent God's anger and punishment of sinners, a perspective confirmed by the citation across the top of the woodcut (Jer. 12: 9). Further confirmation is provided by the citations on the lynx's right (Zeph. 3: 8) and beside the dragon (Ecclus. 43: 23). On the lynx's left, 'Nisi conuersi fueritis gladium suum vibrauit [sc. vibrabit]: arcum suum tetendit et parauit illum' (Ps. 7: 13) associates the lynx with God, a very traditional link. But the citation accompanying the lion and bear

(Prov. 28: 15) apparently conflicts with the others, by suggesting that the lion and bear represent evil forces ('princeps impius'), not righteous vengeance.

The text explains this anomaly by interpreting the lion and bear as 'mauuais princes', whom God will cause to 'faire la guerre aux renars' as an instrument of his anger (fo. e5^r). The narrator also confirms that the other animals denote divine vengeance upon sinners. More importantly, the text shows the sleeping figure in the woodcut to be the narrator: he is in fact not asleep at this point in the text, but the woodcut's suggestion that he is dreaming perhaps gives his vision a prophetic character, since this quality was traditionally associated with certain dreams.[14] In contrast to the other cuts, this image does not represent a scene viewed through the narrator's eyes: the viewpoints of reader and narrator have diverged.

The conjunction of text and image reveals why this is so. The lynx in the woodcut is looking out of the page, at the reader; in the text, however, the narrator says that the lynx 'regardoit assez effrayement la contenance des dessusditz renars' (fo. e5^r). The implication is clear: we, the readers, are foxes. The narrator never says so; this connection must be made through the reader's active interpretative activity, encouraged by the text–image relationship. The divergence of the reader's viewpoint from that of the narrator makes clear what has been latent throughout the text, that Bouchet's readers are themselves foxes, and must therefore heed his lessons.

Clearly, the *Regnars'* woodcuts interact flexibly with various textual elements. Citations, which themselves necessitate metaphorical reading, explain or nuance images; titles give clear or provocative interpretative guidelines; the chapters confirm or revise readers' conclusions. Far from being mere ornament, the woodcuts achieve effects which the text alone cannot. While the visions in the text are mediated by the narrator, the cuts apparently reproduce each vision as the narrator sees it (until chapter 13). Readers, then, are invited to relive the narrator's experience, by being directly confronted with scenes which they must interpret themselves. Hence the woodcuts draw readers' attention to Bouchet's fox metaphor, and encourage them to compare their analyses with the narrator's; in the event, the text exceeds the

[14] See Macrobius, *Commentarii in somnium Scipioni libri duo*, ed. L. Scarpa (Padua, 1981), 1. 3. 1–20.

images, through factual discrepancies, extra detail, or unexpected interpretation, and forces their re-evaluation.

Consequently, the text–image relationship fosters the most crucial activity in interpreting a metaphor; establishing the link between the vehicle (foxes) and the topic domain (human society).[15] What do foxes and people have in common? What does it mean to view human activity as if it were foxes' behaviour? Readers do not passively absorb Bouchet's lessons; they must constantly check the images against the text, both accumulating experience and reassessing it.[16]

The woodcuts thus help foreground Bouchet's elaboration of his metaphor. This is the most prominent self-conscious aspect of the *Regnars*: Bouchet's versification is not especially sophisticated, and although he employs highly rhetorical prose which borrows extensively from Alain Chartier,[17] the printed text's presentation does not highlight these features. The *Regnars* does allude to its own composition in chapter 2, where the authorial persona adopts his distinctive pseudonym and explains his reasons for writing: 'moy qui pour ce petit oeuure ne me oseroye nommer fors le trauerse^{r} des voyes perilleuses meu de pitie et cõpassion seulement apres la reuolucion de plusie^{r}s liures approuuez desquelz ie me suis ayde . . . iay soubz la correction des sages entreprins declairer les figures dessus touchees' (fo. a4^{v}). He states that he is writing primarily in prose 'par ce quelle est de plus facil engin/ et q̃ par icelle on peut mieulx au lõg escripre ce que le sens ordonne' (fo. a4^{v}). However, such metadiscourse is less prominent than the fox metaphors as a means of revealing Bouchet's writerly investment in the text. This may be due to Vérard's appropriation of the *Regnars*: the publisher perhaps excised prefatory material as well as Bouchet's name. Yet despite Vérard's practices, Bouchet is credited in the *Regnars*, in a closing acrostic poem 'ou par les premieres lettres des lignes trouuerez le nom de lacteur de ce present liure et le lieu de sa natiuite' (fo. f2^{v}).[18] Nevertheless, the self-consciousness of the printed *Regnars* consists primarily in its striking fox metaphor, and in the narrator's references to the images

[15] I use the terminology of E. F. Kittay, *Metaphor: Its Cognitive Force and Linguistic Structure* (Oxford, 1987); cf. esp. 140.

[16] Russell, *Emblematic Structures*, 167, suggests that emblems elicited a similar active reading process.

[17] Britnell, *Jean Bouchet*, 84–6; Brown, *Poets, Patrons, and Printers*, 30–2.

[18] Reproduced in Brown, *Poets, Patrons, and Printers*, 259–60.

which foreground the text's 'bookhood'.[19] Indeed, the woodcuts' importance is suggested by their close reproduction in later editions. Vérard's cuts are trimmed for reuse in his second, quarto edition.[20] Michel Le Noir's edition crudely copies Vérard's images, with only minor differences in detail, and employs the same layout; the 1522 and 1530 editions reuse these copies.

Bouchet later reworked the *Regnars* in a version which survives as Poitiers, Médiathèque François-Mitterrand, ms. 440; this manuscript dates from 1531, though the revision itself is probably earlier.[21] In view of Bouchet's tendency to print revised versions of his work, it is difficult to explain why this text remained unpublished. Disappointingly, the manuscript contains only two illustrations, though spaces have been left for others. One of these images, discussed below, is new; the other (fo. D4^r^) copies the woodcut accompanying chapter 5 in the printed *Regnars*. It must remain uncertain whether the other fox images, if they had been provided, would similarly have resembled those of the printed version.

Bouchet's revised text differs from the printed *Regnars* in various significant ways, reflected in the manuscript's presentation. A preface and two chapters are added, other sections expanded, and the sequence of chapters renumbered to begin with the first vision (chapter 4 in the printed text). Moreover, by altering the visions, Bouchet clarifies the fox metaphors. The *Declaration* explains the fox image more fully; the authorial persona claims he has presented humans as foxes and wolves because humans who fail to use their reason are no better than animals: 'sõmes mis audessoubz de nature humaine et sõmes faiz semblẽs par transformacion poeticque aux bestes irraisonnables' (fo. B2^{r}). Humans are now compared to both foxes and wolves (fos. B4^{v}–8^{r}), though this does not nuance the metaphor: wolves' traits are in practice

[19] I borrow this term from Couturier, *Textual Communication*, ch. 2.

[20] See Mortimer, *Harvard*, i. 142–4. The cuts for chs. 5 and 12 are trimmed further for reuse in Robert Gobin's *Les Loups ravissans*, Paris, Anthoine Vérard, n.d. (BNF, Rés. Ye 851), fos. l5^{v}, b2^{v}.

[21] Britnell, *Jean Bouchet*, 88. A. Richard, 'Notes biographiques sur les Bouchet, imprimeurs et procureurs à Poitiers au XVIe siècle', 556–61, describes the volume, though not entirely accurately. Overlooking Bouchet's alterations to the sequence of chapters, Richard mistakenly claims that a chapter has been omitted from the manuscript version. He suggests that corrections to the text may be in Bouchet's hand, and that the manuscript possibly constituted a response to Philippe Le Noir's reprinting of the *Regnars* in 1531.

indistinguishable from those of foxes. Chapter 3, new to the manuscript revision, presents a vision of wolves in sheep's clothing holding pickaxes and keys outside a closed city (fos. E6^{v}–G5^{r}). The narrator initially interprets these as 'des pasteurs de leglise', but a Latin legend above the city suggests the correct reading, that the wolves represent 'les mauluaiz gouuerneurs des villes et citez/ et ceulx qui ont la police et salut publicque entre mains' (fo. E7^{r}). His initial misreading encourages readers to concentrate upon the image's appropriateness, further reinforcing the visions' tendency to thematize the fox/wolf metaphor.

Chapter 6 of the printed *Regnars*, renumbered chapter 5 in the reworking, simplifies the image of foxes at bay: it focuses solely on 'folz esperans', omitting all reference to *joncheurs* (fos. K4^{v}–M7^{v}). A long discussion of free will is added to the material on foolish hope, giving it an orthodox theological foundation, and various exempla of the transience of worldly goods are cited. Other chapters similarly contain more exempla than the printed versions, and expound metaphors' meanings more clearly. Hence the fox metaphor retains its importance, but loses some of its sophistication as orthodox moral teaching attains greater relative prominence.

This tendency is counterbalanced by the way in which the reworking presents the authorial persona in more self-conscious terms. Bouchet's activity is most strikingly thematized in a preface, unique to the reworking:

> Ie Iehan bouchet de poictiers qui en ce liure me nõme le trauerseur des voies perilleuses Apres la premiere traduction de latin en francois de la nef des folz/ ou pour entree ie occupay mon petit entendement/ Enuye des abuz du monde/ et mesprisant la facon de viure des mondains en tous estatz/ deliberay en escrire ma fãtasie/ Et considerant que a ce me pourroient grandement seruir vingt et huit vers et metres elegãment composez par ce notable docteur mesßr Sebastian brand/[22] Qui premier auoit compose en langue theutonique lad' nef des foulz/ Sur iceulx vers et metres/ pour contenter et satisfaire a mon entreprise/ . . . ie cõmencay soubz la conduicte ⁊ inspiracion diuine rediger par escript mes fantasies par vng petit traicte intitulle le liure des renars/ esperant le paracheuer/ ce qui ne me fut lors possible/ Et pource q̃ ie trouuay lẹs loups pouoir seruir a ma matiere/ voulu les y emploier/ et du tout redigeay par escript sept ou huit cayers/ desq̃lz [*sic*] auoient depuis este transportez ꝑ moiens ne scay quelz entre

[22] i.e. *De spectaculo*, which follows the preface on fo. A2^{r-v}.

les mains daucuns libraires/ qui ont iceulx imprimez/ et impose tiltre et acteur a leur plaisir.

Et voiant que iestois deceu de mon entencion/ et que ce petit euure nestoit paracheue/ et y auoit plusieurs faultes et obmissions. . . . desirant de tout mon cueur mectre fin a la chose cõmancee/ iay reueu corrige et addicionne de plus de moictie led' euure en rudes termes ⁊ langage vulgaire. (fo. A1^{r-v})

Bouchet recounts his text's history, from its genesis through its unauthorized publication to the present revision. Whether the details be true or not, Bouchet invites readers to trace his work as poet through the different stages of the piece's elaboration, and to identify him, the historical author, with the Traverseur, the homodiegetic witness-narrator.

This conflation of author and narrator is further encouraged by a second additional chapter, chapter 4, entitled 'Le debat dauarice et de pauurete qui parle en ꝑsõne de lacteur' (fos. G5^{v}–K4^{r}). Here, the narrator is seized by wolves and brought before Avarice, who criticizes his attacks on her: 'semble a veoir tes escriptz que chũn me deuroie fuyr' (fo. G7^{r}). The *acteur* defends his writing as having a moral purpose; he and Avarice then discuss the various kinds of poverty. Although wolves are present in this chapter, they are peripheral to the central debate: Bouchet's concerns as author take precedence over his didactic metaphor.[23] Not only are author and narrator collapsed into each other, but the boundaries of the narration are metaleptically blurred: Avarice, a character within the diegesis, steps out of this frame to criticize Bouchet's writing. Furthermore, the chapter title suggests that an allegorical figure, Pauvreté, assumes Bouchet's identity; the authorial persona thus becomes a personification, existing on the same ontological level as Avarice. Although this transformation is signalled only in the title, and not made explicit in the text, it underlines the interpenetration of diegetic levels. The text becomes strikingly reflexive as Bouchet incorporates a defence of his writerly practice into the fiction he narrates.

The new illustration in the manuscript reinforces the reworking's thematization of Bouchet's activity (fo. A3^{r}; Plate 5.4). It depicts a man surrounded by foxes, though, since the accompanying scroll is blank, it is uncertain whether the man is Bouchet or the Traverseur. Since the two are conflated within the text, this

[23] Bouchet also defends poetry in the *Temple de Bonne Renommee*, ed. G. Bellati (Milan, 1992), 319–26, vv. 3319–608.

PLATE 5.4 Bouchet, or the Traverseur, and foxes: Poitiers, Médiathèque François-Mitterrand, *Les Regnars traversant*, ms. 440, fo. A3^{r}

ambiguity is all the more suggestive. The retention of the closing acrostic poem confirms Bouchet's prominence; this time the acrostic is highlighted through spacing (fo. BB3r; Plate 5.5).

A final important revision to the *Regnars* consists of the copious marginal annotations which now flank the text. Visually distinguished by a smaller script and underlining, they are citations from learned texts, especially Boethius, Cicero, and Ecclesiastes, which endow Bouchet's moral arguments with a Latinate authority. Such annotations characterize Bouchet's moral writing from 1512 to the mid-1520s. The margins of many more of his works contain references to learned sources, or summarizing passages which clarify the way in which Bouchet articulates his argument in didactic or historical texts.[24] In the revised *Regnars*, the citations sometimes pursue points further than the text: thus the *acteur* claims in chapter 1 that he cannot express the many 'publicques pechez du peuple' (fo. D1v), but a lengthy annotation (Isa. 59: 1–4) accuses the audience of various sins. However, the annotations' primary effect is to persuade readers of Bouchet's orthodoxy and erudition, thereby helping secure his reputation for authoritative didactic writing.

The Poitiers manuscript, then, places more stress than the printed *Regnars* on Bouchet's thematization of writing, his scholarly authority, and the way in which he invites identification of himself with the narrator. The volume's annotations, paratext, and illustration all highlight these new tendencies. Though metaphor remains important, it is not the dominant self-conscious technique as in the earlier version. Because of the annotations' presence, the manuscript evinces more visual variety than the editions: besides the illustrations, and the different shapes on the page of verse and prose, it also presents blocks of text in the margins, in a distinctive script. This not only underlines the variety of expressive resources which Bouchet uses (images, prose, verse, learned sources), but also gives readers an even stronger impression of the work's 'bookhood' than does the printed text. While Bouchet's writerly activity is made prominent, it is equally clear that what he has produced is a material object.

[24] For examples of annotations and their use, see Bouchet, *Déploration*, 24–6, 35–6, and my discussion of the *Labirynth de Fortune* below. Both summaries and source references appear, among numerous other texts, in the *Triumphes de la noble et amoureuse dame*; see Britnell, *Jean Bouchet*, ch. 7, esp. 230–5.

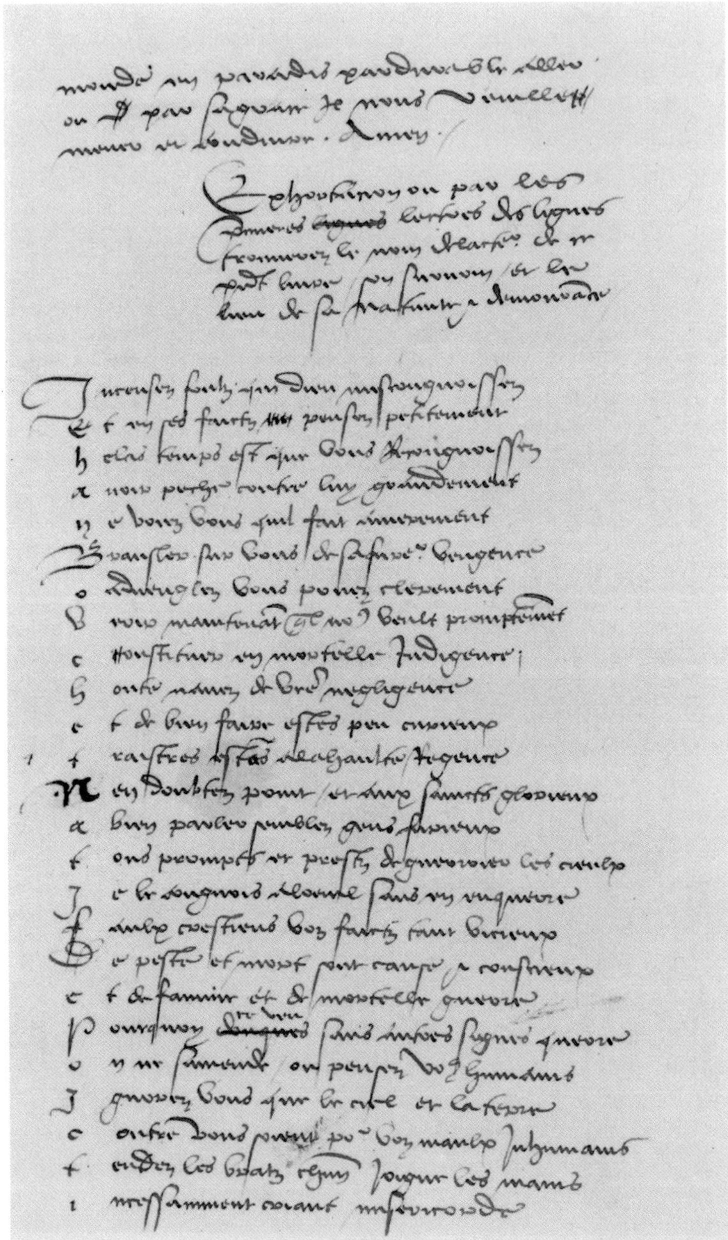

PLATE 5.5 The closing acrostic: Poitiers, Médiathèque François-Mitterrand, *Les Regnars traversant*, ms. 440, fo. BB3[r]

Le Labirynth de Fortune

The *Labirynth de Fortune* typifies Bouchet's moral and theological writing in the 1520s and 1530s. Its first two editions appeared in Poitiers; three further editions were published in Paris.[25] A sophisticated text incorporating a wide range of Latinate erudition from both classical and Christian writers, the *Labirynth* primarily addresses questions of free will and fate.[26] It opens with the *acteur*, explicitly named as Jean, learning of the death of Artus Gouffier, *grand-maître* of France; a figure identified as Voix Discrete laments Gouffier in verse (fos. A1^{v}–A6^{r}). The *acteur* is led to reflect on the mutability of Fortune, which he considers in his spare time over a period of two years. Then, in August 1521 when he and his family had left Poitiers to escape an outbreak of plague, he dreams that he meets the personification Suggestion, followed by eleven groups of people clad in symbolic colours which reflect their preoccupations. He introduces himself, and joins those dressed in grey, who wish to obtain good fortune through hard work. Entering the Labyrinth of Fortune, he meets Humaine Discipline, an allegory of classical and humanistic learning, who shows him Fortune on her throne. The latter is represented in traditional guise:[27] half-white and half-black, blind and deaf, she is accompanied by Eur and Maleur, who proceed to argue about their relative importance.

Humaine Discipline informs Jean that the maze 'nest riẽs fors lestat de cestuy mõde' (fo. E6^{r}): people of all types wander within it, encountering unexpected misfortune and swift change. However, Jean rejects his guide's conception of Fortune, raising orthodox Christian objections to the view that Fortune is a goddess. His criticism is reinforced by the arrival of Veritable Doctrine, who personifies Christian teaching and vituperates against the vanity of art and knowledge, especially poetry. She engages in a debate with Humaine Discipline, after which Jean wakes from

[25] Poitiers, Enguilbert de Marnef and Jacques Bouchet, n.d. [*c.*1522] (Musée Condé, IV, E. 82); Enguilbert de Marnef and Jacques Bouchet, 26 March 1524 (Arsenal, 8° BL 8744); Paris, Alain Lotrian, n.d. (BNF, Rés. Ye 353); Philippe Le Noir, n.d. (two editions, of which representative copies are Bodleian, Douce BB 507, and BNF, Rés. Ye 354). See Britnell, *Jean Bouchet*, 311. I quote from the second edition throughout.

[26] See Britnell, *Jean Bouchet*, esp. 48–9, 197–8, 205–8.

[27] See H. R. Patch, *The Goddess Fortuna in Mediaeval Literature* (Cambridge, Mass, 1927), 40–9.

his dream to find Veritable Doctrine beside him. When Jean asks why human fortunes vary, she replies with an orthodox exposition of such doctrines as first and second causes, celestial influences, and free will. The goddess Fortune, she claims, is a mere poetic fiction; the labyrinth is a valid metaphor for life because it demonstrates that 'permanence en ce monde ny a' (fo. K6^{r}), but Humaine Discipline is wrong in attributing this mutability to Fortune, Eur, and Maleur. She closes the first part of the text by asserting that faith, hope, and charity are needed as guides in the Labyrinth.

The second part opens with Jean asking where to find these theological virtues; he is told that they are in 'le seiour des troys dames' (fo. M7^{v}), which must be reached by a steep, rocky path, and given Vertu as a guide. Not without difficulty, Jean arrives at the Sejour to find the virtues, who each deliver long addresses on theological issues. Finally, the various humans who have reached the Sejour divide up into those who follow the active life (who return to the Labyrinth, guided by the four cardinal virtues) and those opting for the contemplative life (who remain in the Sejour, and are given wings so that they can travel to the Labyrinth with spiritual nourishment for their active fellows).

The *Labirynth* is a self-conscious text in various respects. First, the labyrinth image is not only a metaphor for the world under Fortune's dominion, but is replicated and reworked throughout the text in manifold guises, producing a thematic and formal density which foregrounds Bouchet's compositional artistry. The complex of imagery is best addressed through the description of the Labyrinth's interior:

> il sembloit vng iardrin [*sic*] plain de allees
> Larges assez ⁊ fort bien esgallees
> Dont y auoit plus de mil et cinq cens/
> Et si estoit [*sic*] composees dun tel sens
> Que dune allee en lautre on se destourne
> Et ne veoit on en quel estat on tourne
> Et tellement que a peine on peut sortir
> Sans de quelcun (qui lentend) se assortir/
> Car en cuidant aller par vne voie
> Et vng cartier fort loing on se desuoye. (fo. B4^{r})

Distractions abound: games, feasts, women, and pastoral scenes can be found, while various vices lead wanderers astray.

Humaine Discipline interprets the labyrinth as an image of earthly life, which reflects the world's bewildering variety. It is not this reading which Veritable Doctrine contests, but the association of the labyrinth with Fortune; indeed, the maze's very structure suggests that Humaine Discipline is mistaken. Penelope Reed Doob distinguishes two types of labyrinth: the multicursal type with forking paths, and the unicursal variety with a single twisted path. In antique and medieval literature, each type had consistent metaphorical overtones. Multicursal mazes connoted moral choice, the wanderer's responsibility for choosing his path, and the potential for transcending the maze's confusion through experience or revelation; unicursal mazes implied a fatalistic view of life due to the lack of choice on the part of the wanderer, who, subject to the path's vagaries, required not proper judgement in making correct choices but perseverance in following the path to its end.[28] Bouchet's Labyrinth is clearly multicursal. Its many paths and distractions lead wanderers into error; moreover, it is described as having multiple entrances: there are 'vnze portaulx', one for each group of Suggestion's followers (fo. B3r). Yet Humaine Discipline's conception of life as being ruled by Fortune is more applicable to *unicursal* labyrinths: she has apparently misinterpreted the maze. In contrast, Veritable Doctrine's Christian view of life, which gives due weight to both free will and Providence, is more appropriate to a multicursal labyrinth. Veritable Doctrine thus supplants the pagan reading of the Labyrinth with one better attested by the tradition of maze imagery.

The second part of the *Labirynth* introduces another maze-like edifice: the Sejour, the world of the spirit, revealed only by Christian theology. It is reached by a difficult path which, like that in a unicursal labyrinth, requires perseverance: Jean is hindered when a 'dur chillou de regret amoureux' hurts his foot (fo. M8v), and Congnoissance de Soy must come to his aid. The latter expounds the familiar didactic topos of the contrasting paths of vice and virtue, implying the necessity of choosing one's path properly (fo. N8r). The Sejour itself, an enclosed garden, is described in terms strongly reminiscent of the Labyrinth:

[28] P. R. Doob, *The Idea of the Labyrinth from Classical Antiquity through the Middle Ages* (Ithaca, NY, 1990), 3, 46–51. Much of the following discussion of mazes in the *Labirynth* is indebted to Doob's reflections.

tout autour y auoit grans allees
Que theologie auoit bien esgallees/
Aux quattre coings tourelles y auoit
Faictes darbriers dont grant odeur sortoit
En lune vy vng homme/ en lautre vng ange
Vng veau en lautre/ en lautre vng aigle extrãge
Me fut cela/ car tous auoient des yeulx
Helles aussi pour voller iuc es cieulx
Et dauant eulx auoient chascun vng liure/
Qui tesmoignage a mon petit sens liure
Que cestoient ceulx lesquelx auoient escript
Les nobles faictz du benoist ihesuchrist. (fo. O1^{r})

The various 'allees' suggest a labyrinthine structure, while the repetition of the *allees/esgallees* rhyme from the Labyrinth's description links the two gardens, highlighting the way in which the Labyrinth and Sejour complement each other as the material and spiritual worlds. Further parallels include the great variety of pilgrims Jean sees on the road to the Sejour (fo. M8^{r}), echoing the groups inside the Labyrinth, and Jean's need for a guide, this time Vertu rather than Humaine Discipline. The Evangelists at the corners of the Sejour recall the Opinions which guard the Labyrinth's portals, admitting humans selectively. Yet, unlike the Opinions, the Evangelists suggest the possibility of transcending the labyrinth, through flight. The contrast between the mazes reflects that between the material and spiritual worlds: the Sejour does not suggest confusion, though the path leading to it may deter some travellers. Bouchet's text thus dramatizes its didactic purport both through the *acteur*'s progression towards knowledge, via the precepts of the personifications which address him, and through the labyrinth image's evolution. Although the original Labyrinth reappears at the end of the narrative, there is a clear movement from bewilderment to clarity, common in literary texts centred round labyrinths.[29]

This metaphorical intricacy alone would render the *Labirynth* self-conscious, in that Bouchet foregrounds his use of the image. However, the labyrinth functions as an image not only of earthly life, but also of Bouchet's text. In several important respects, the *Labirynth* displays its own 'labyrinthicity'.[30] All labyrinths are char-

[29] Doob, *Idea*, 52. On the Sejour, see also Cowling, *Building*, 135–6.
[30] The term is coined in Doob, *Idea*, 2.

acterized by the presence of tortuous *ambages* and a quality of planned chaos,[31] which Bouchet works into his text in various ways. There is, for instance, a striking multiplicity of voices: the *acteur*; the writers of liminary material, Bouchet himself, Antoine Ardillon, and Jean d'Auton;[32] Humaine Discipline, Veritable Doctrine, Voix Discrete, Eur, Maleur, the theological virtues, and so forth. More voices are present in the form of Latin marginal annotations from classical and Christian texts, which provide the sources for the arguments Bouchet expounds in narrative form. Formal variety underpins this polyphony: Bouchet employs a wide range of stanza forms, varying for each personification.[33] There is a strong impression of both disorientating variety and impressive artistry, typical of both real and literary labyrinths.

Moreover, the Labyrinth is itself rendered literary, by being related to poetry. Veritable Doctrine tells Jean that Fortune is a

> simple figure
> Des poetiqz/ lesquelz nont point congneu
> Que dieu puissant a sagement pourueu
> A ce qui est es cieulx/ ꝛ en la terre. (fo. K5^{v})

Though this does not apply to the Labyrinth itself, the characterization of Fortune as a mere pagan metaphor underlines how the Labyrinth similarly is not real, but rather a metaphor for earthly life. The Labyrinth is also assimilated to Jean's dream: it has ivory portals, a topos in literature about dreams (fo. B3^{r}). Yet Bouchet presents dream-visions as essentially literary devices, whose only truth lies in their utility as metaphors. Both Jean and Veritable Doctrine criticize dreams' epistemological validity: indeed, dreams associated with ivory gates were traditionally false, according to interpretative conventions established by Macrobius.[34] The Labyrinth's association with dreams thus confirms its status as literature.

Most significantly, however, Bouchet's own practice may be regarded as labyrinthine. In composing a complex work of art, he plays the role of maze-builder; in producing a handy synthesis of learned texts, he is an Ariadne-figure, guiding us through a maze

[31] Ibid. 51–5.
[32] On all Bouchet's correspondents, see Britnell, *Jean Bouchet*, app. A.
[33] Ibid. 32–3.
[34] Macrobius, *Commentarii*, I. 3. 17. See also F. Cornilliat, 'Aspects du songe chez les derniers rhétoriqueurs: Analyse du *Labirynth de Fortune* et du *Séjour d'Honneur*', 22–7.

of erudition; yet he himself has walked this latter maze by assimilating the texts from which he cites, and his homodiegetic surrogate Jean must negotiate the Labyrinth and Sejour. Hence Bouchet may be regarded as simultaneously within and outside the maze, both experiencing its *ambages* and constructing them anew. The labyrinth image, then, embraces Bouchet's compositional and compilatory practice, as well as the discussion of free will and the moral topos of uncertain earthly existence. Its prominence and far-reaching implications ensure that readers are aware of Bouchet's writerly investment in the *Labirynth*, not only as technical artistry or as a reinforcement of his arguments, but also as a tacit thematization of poetic activity.

Bouchet also thematizes poetry, especially his own, in other ways. Fortune is presented from the beginning as a poetic construct, 'Faiz et tissuz par dame poesie' (fo. B6^r), while Veritable Doctrine criticizes not only poets who regard Fortune as a goddess, but all mythological fictions:

> En poesie il nya que mensonge
> Poetes sont communement menteurs
> Tous leurs ditz sont fables/ fictions/ songes
> De nouueaulx cas sont tousiours inuenteurs
> Ou lon peut prandre et cueillir grans erreurs
> Ientends les gens lesquelz nont la science
> Lart/ le scauoir/ lesprit/ lesperience/
> Ces fictions a droict moraliser. (fo. F3^r)

Readers are thus invited to go beyond Bouchet's surface narrative and extract theological and moral lessons. Yet, as has been shown, the *Labirynth* displays a notable degree of 'poeticity':[35] its surface cannot simply be ignored. The *Labirynth* does not merely restate Christian truths in French, despite Bouchet's protestations that he has written it

> seulement pour a tous remonstrer
> Commant il fault vertueux se monstrer. (fo. V4^r)

In the light of Veritable Doctrine's critique, Bouchet can be seen to be capable both of writing mythological fictions and of moralizing them effectively.

Furthermore, readers are often tempted to identify the *acteur*

[35] See R. Jakobson, 'What Is Poetry?', 174.

with Bouchet as historical author. The early references to plague-stricken Poitiers are followed by Jean's meeting with Suggestion:

> ie luy dis que le nom ie portoye
> De Iehan bouchet/ demourant a poictiers. (fo. B2^{r})

Jean's affliction by the 'dur chillou' on the way to the Sejour exemplifies the humorous, self-deprecating presentation of first-person narrators characteristic of Bouchet's later narratives.[36] Consequently, the *Labirynth* frequently draws readers' attention to the activity of poets, and particularly of Bouchet, whom the *acteur* seems to represent directly. The poem foregrounds its status as a constructed literary artefact.

The *Labirynth*'s editions do not always convey its self-conscious aspects. Though each edition's visual presentation differs from that of the others, their physical form can be compared in four principal areas: illustration, marginal annotations, liminary material, and layout. I shall examine these areas in this order, and demonstrate that while the Poitiers editions tend to respect the text's labyrinthine traits, Paris publishers prefer to highlight traditional moral themes.

The Poitiers editions bear a large woodcut depicting Daedalus and Icarus escaping the Cretan labyrinth, represented simultaneously in planar form and as an external view (fo. $^{\pi}$A8^{v}; Plate 5.6). The Labyrinth of Fortune, then, is not itself represented; rather, the image is a generic one permitting easy recognition of the maze motif. However, attentive readers may identify significant textual parallels. The use of wings to escape the Cretan labyrinth echoes the wings given to those who choose the contemplative life at the end of Bouchet's narrative, permitting both their eventual ascent to Paradise and their transcendence of earthly *ambages* while they live. Bouchet's moral aims as well as his labyrinth image are thereby underlined. A similar effect is achieved by the juxtaposition of planar and three-dimensional representations, which reflects the different readings of the Labyrinth in the text. The three-dimensional depiction renders the maze mysterious and threatening, since it does not permit an overview; the plan demystifies the labyrinth's *ambages* and permits a route through it to be plotted. The former representation corresponds to Humaine

[36] See Britnell, *Jean Bouchet*, 51–2.

PLATE 5.6 The Cretan labyrinth in the second Poitiers *Labirynth de Fortune*: Poitiers, Médiathèque François-Mitterrand, *Le Labirynth de Fortune*, DM. 9, fo. $^{\pi}$A8^{v}

Discipline's pagan obscurantism which denies humans freedom of action, and the latter to Veritable Doctrine's perspective: able to transcend and look down upon the world by virtue of the flight which faith permits her, she can see earthly life for what it truly is, a part of God's overall plan within which one can exercise free will and choice.[37] Hence the labyrinth's apparently confusing dual representation is explicable in terms of the two readings Bouchet offers; again, readerly attention is drawn to the metaphor's implications.

The Paris editions all employ stock woodcuts, rather than labyrinth images. Lotrian's edition bears a cut of a teacher and pupils on its title-page, foregrounding the *Labirynth*'s didactic value, and an image on the final page (fo. GG6^{v}) from Pierre Gringore's *Fantasies de Mere Sote*, depicting an emperor discovering a sarcophagus engraved with legends such as 'IAY DESPENDV' or 'IE LAY PERDV'.[38] In the *Fantasies*, this illustrated an exemplum of the benefits of good advice; however, in the context of the *Labirynth* the legends imply the transience of worldly goods under Fortune's dominion, rather than issues of free will and fate. By using a stock image of tangential relevance to the text, Lotrian highlights the traditional notion of Fortune, at the expense of both the poem's theological preoccupations and its governing metaphor. The same applies to Philippe Le Noir's editions. One of these bears a Wheel of Fortune image, probably reused (BNF Rés. Ye 354, fo. hh8^{r}); the other has a stock cut of a king and courtiers on its title-page.[39]

The annotations consist of references to or quotations from Bouchet's various source texts, overwhelmingly in Latin. A wide range of sources is represented, often discussing thorny issues more fully than the *Labirynth* itself.[40] Whether or not Bouchet's first readers could read Latin, and therefore appreciate the finer learned points confined to the marginalia, the very use of Latin secures a culturally based authority for his text. The Poitiers

[37] It should not be assumed that, because the maze in the plan is primarily unicursal, it is more appropriate to a fatalistic than to a Christian doctrine. Unicursal representations were practically universal in visual art (Doob, *Idea*, ch. 5); the woodcut owes more in this respect to artistic convention than to Bouchet's text.

[38] These images are reproduced in A. Tchemerzine, *Bibliographie d'éditions originales et rares d'auteurs français des XVe, XVIe, XVIIe et XVIIIe siècles* (2nd edn. Paris, 1977), ii. 33. See also P. Gringore, *Les Fantasies de Mere Sote*, ed. R. L. Frautschi (Chapel Hill, NC, 1962), woodcut 15.

[39] Reproduced in Tchemerzine, *Bibliographie*, ii. 32, 34.

[40] See Britnell, *Jean Bouchet*, 206–7.

editions visually reinforce this authority by presenting the annotations in a small rotunda typeface, which has a more Latinate flavour than the type used for Bouchet's verse, a bastard of the kind primarily employed for vernacular texts (fo. G1ʳ; Plate 5.7). Moreover, the variation of type accentuates the way in which the annotations increase the *Labirynth*'s formal variety by complementing French verse with Latin verse and prose.

This increased variety has further effects. The density of annotation gives an instant clue as to whether or not passages are doctrinally significant: the description of Suggestion and her followers lacks marginalia, and can therefore be read in a more relaxed way than heavily annotated discussions of virtue. Consequently, the degree of annotation at each point in the text helps signal what readers should be privileging, the narrative surface or the doctrinal lessons. This switching of priorities is reinforced by the constant switches of focus which the marginalia demand of readers. Our eyes are drawn from the verse to the annotations and back again, maybe even to other books in order to look up references. Text and annotation are competing attractions; indeed, by forcing readers to oscillate in this fashion, the marginalia turn every page on which they appear into a kind of labyrinth. Whenever an annotation appears, readers are faced with a *bivium*: we must decide either to continue reading the verse or to follow the *ambages* of the annotations. While this is always the case in a text with marginal glosses, the text–annotation relationship attains a specific value in the *Labirynth*: because the equivalence between text and maze is so prominent, the metaphorical significance of the *ambages* on the page is fully actualized. In addition, the constant switching of readers' preoccupations results in a dual reaction to the *Labirynth*, which reflects a characteristic feature of mazes.[41] According to the reader's perspective at any one time, the poem may be either a coherent, 'unicursal' allegorical narrative or a 'multicursal' compilation of learned texts. Readers must transcend the surface narrative to extract the appropriate lessons, just as the contemplative life permits transcendence of the Labyrinth in the poem, but can never ignore the surface entirely (even the contemplative must live in the world). Bouchet's marginalia thus reinforce, nuance, and complicate the labyrinth metaphor, until it embraces not only the

[41] See Doob, *Idea*, *passim*.

Bien temporelz/et dor les plaines tables
Lesquelz ne sont perpetuelz ne stables
Aux possesseurs ilz sont des tormens maints

Nous cougnoissons les changemēs soudains
De bien en mal que ont les pauures mondains
Des biens du monde ainsi souuent muables
Non quant a dieu
Cas que disons fortuitz sont certains
Tousiours a dieu/mais no⁹ qui sōmes plains
Dobscuritez et choses decepuables
Quant nous voions telz biens si peu durables
Nous les disons de fortune prochains.
Non quant a dieu
Humaine discipline
En biens mondains vous verrez prosperer
Les grans pecheurs/ou lon deust esperer
Rabaissement de leur grosse richesse
Et auoir bruyt sur ceulx qui ont sagesse
Dont les raisons ie ne puis digerer

Et si voulez es bons considerer
Deulx ne pouez que indigence inferer
Tousiours maleur les moleste et oppresse
En biens mondains
Ceulx que fortune entend accelerer
Es grans honneurs/ne les fait ponderer
Ne bien ne mal/mais au gaings les adresse
Sans traueiller/les aultres ont destresse
Riches sont ceulx esquelz veult adherer
En biens mondains
Veritable doctrine
Qui est de dieu par raison ponderee
Ne tasche auoir richesse immoderee

Licet aliqua sint fortuita respectu hominum nihil tamen respectu dei q̄ oia preordinat & preuidet a quo fit effectus fortuitus. Btūs tho. cōtra gē=ti. li. iij. c. lxxxix.

Subibat q̄ me non de nihilo veteris prisceq̄ doctrine viros finxisse ac pronunciasse cecam & prorsus exoculatā inesse fortunā que semper suas opes ad malos et indignos cōferat. apuleius li. vij.

Qui volūt diuites fieri incidūt; in tentationē & laqueum diaboli & desideria multa inutilia & no=

G i

PLATE 5.7 Typographic variety in the second Poitiers *Labirynth*: Poitiers, Médiathèque François-Mitterrand, *Le Labirynth de Fortune*, DM. 9, fo. G1[r]

text's striking allegory, theological arguments, and formal sophistication, but also the book as object.

Like many of Bouchet's later works, the *Labirynth* is accompanied by various liminary pieces. These comprise: a privilege to Enguilbert de Marnef; a Latin letter to Bouchet from Antoine Ardillon; a reply in Latin by Bouchet (these first three texts appear in the Poitiers editions only); a table of contents; a letter from Bouchet to Marguerite de Navarre; then, following the *Labirynth* proper, a letter from Bouchet to Jacques Prévost, a theologian who had checked the text's orthodoxy, and an exchange of verse epistles between Bouchet and Jean d'Auton. The proliferation of these texts increases the *Labirynth*'s formal and linguistic variety, and its plurality of voices. As does the preface in the revised *Regnars*, the epistolary material often anchors Bouchet in the *hors-texte* and encourages identification of the first-person authorial voice with the historical author. This is especially striking in the epistle to d'Auton, in which Bouchet reviews his career (fos. V7^{r}–x1^{v}). He claims to have composed his poetry in the evenings, his days being occupied with judicial business (fo. V7^{v}); criticizing his detractors, he protests his orthodoxy:

> Ie nescriz rien qui ne soit approuue
> Et par histoire ou bon acteur prouue. (fo. V7^{v})

These autobiographical details both advertise Bouchet's compositional investment in the text and secure his reputation as a poet. Ardillon's and d'Auton's pieces, which praise Bouchet lavishly, complement Bouchet's self-justification by providing useful publicity, and guaranteeing his poetic and doctrinal competence (fos. $^{\pi}$A2^{r}, x2^{r}–4^{v}).[42]

The liminary texts effectively reflect the balance between moral lessons and complex rhetorical organization in the *Labirynth*. Ardillon focuses upon the text's didactic value: 'vt lectores a vicijs auocare: dissuadere: deterrere: contra vero ad virtutem excitare:

[42] Such promotional discourse is common in the paratext of Bouchet's work. A significant example is the *Triumphes de la noble et amoureuse dame*, which narrates the soul's journey through life via an allegorical tale replete with religious teaching (Britnell, *Jean Bouchet*, ch. 7). Copious liminary material in its *editio princeps* delineates and justifies Bouchet's enterprise: verse epistles to Eleanor of Austria and Louis de Ronsard, a prose 'Epistre de lacteur', and a 'Prelude' in verse are complemented by certificates from theologians which guarantee the text's orthodoxy (Poitiers, Jacques Bouchet, 20 June 1530; BSG, Rés. Y fol. 141 inv. 201).

adhortari: ac inflamare . . . possit' (fo. $^{\pi}$A2^{r}). Bouchet himself underlines his synthesis of Latin learning, especially in his letter to Prévost, which mentions

> la reserche et labeur que iay prins a extraire les dictz/ oppinions/ τ sentences daucuns nobles poetes/ eloquẽs orateurs/ philosophes naturelz et moraulx/ docteurs saincts et catholicques/ loix ciuilles et canoniques/ coctez en la marge du langage vulgaire. Le tout a lhonneur et gloire de la trinite premierement/ a la recreation de mon esprit/ τ pour en reuoluant les liures apprẽdre a me congnoistre/ τ finablement a la consolation des dames/ τ aultres personnes qui nentendent les lettres latines/ lesquelz pourront veoir en briefue substance/ ce qui est en plusieurs volumes de liures confusement contenu. (fo. V6^{r})

This passage is especially noteworthy for its presentation of Bouchet's activity as synthesist. The service he performs for readers without Latin (or without access to numerous books) underlines his role as Ariadne, helping his readers negotiate the labyrinth of learning in which pagan and Christian authors come to worryingly diverse conclusions regarding fortune.

In his letter to Marguerite, Bouchet summarizes the text, giving equal weight to the narrative framework and to the theological lessons. Moreover, he sets the Labyrinth in a tradition of monuments for illustrious figures,[43] implying that the poem will primarily commemorate Artus Gouffier (fos. $^{\pi}$A5^{v}–6^{v}). This is not the case, but the juxtaposition of the Labyrinth's apparent significance here with its real significance in the poem underlines the polysemy of the maze image, which can support two possible functions, memorial and doctrinal. The labyrinth's ambivalence, and the respect for the narrative framework in Bouchet's plot summary, show that the poem's allegorical fictions cannot be kept separate from its doctrinal lessons, or seen as entirely subordinate to them. Rather, they reflect and reinforce Bouchet's teaching, by elaborating a narrative of enlightenment which turns upon the maze's reinterpretation. By valorizing both aspects of the *Labirynth*, Bouchet ensures that its polyphonic complexity is properly appreciated.

Indeed, Bouchet's epistle to d'Auton augments the density of the poem's imagery by introducing another metaphor for life to compete with the labyrinth allegory. This is the image of life as

[43] Cowling, *Building*, 164–5, notes the self-justificatory nature of these remarks.

theatre, underpinned by the assumption that life, like a play, is unreal:[44]

> ce ne sont de tous estatz que rolles . . .
> Et leur mistere est la mondaine vie. . . .
> Celuy qui porte aux ioueurs le papier
> Se nomme abuz. (fo. V8$^{r\ v}$)

This instance of the *contemptus mundi* topos partly contrasts with the labyrinth motif, which recognizes life's *ambages* but also its value, and partly reinforces it by underlining fortune's deceptiveness. Bouchet demonstrates his ingenuity to his fellow-poet by exploring another metaphor instead of restating the maze image; he again shows that he can do more than compile and translate orthodox precepts.

D'Auton's reply to Bouchet is highly significant for our appreciation of the labyrinth image. The final text in the volume, it explicitly identifies Bouchet's poem with a labyrinth:

> a la tienne requeste
> Iay passe oultre ⁊ me suis mis en queste
> Dentrer dedans pensant trouuer brisee
> Pour me donner saillie auctorisee. (fo. x3^{v})

Shortly afterwards, d'Auton again refers to the text as 'ce riche labyrinthe' (ibid.). If readers have not already appreciated the poem's labyrinthicity, d'Auton's piece draws their attention to it, compelling a retrospective reading. This epistle, then, confirms the importance of Bouchet's labyrinth image as an emblem of the poem's structure.

Hence, in various ways, the liminary pieces reinforce the self-conscious processes I have noted, the rich and prominent labyrinth image and the highlighting of Bouchet's activity as poet. Indeed, they form a labyrinthine pattern of their own, complementing that of the text. Many of the pieces refer to earlier material: Ardillon's text and Bouchet's letter to Marguerite mention the exchanges with Prévost and d'Auton, while Bouchet's pieces to Ardillon and d'Auton mention his letter to Prévost. A complex pattern builds up, in which it is assumed that readers can compare the different pieces: this movement back and forth amid the

44 See E. R. Curtius, *Europäische Literatur und lateinisches Mittelalter* (2nd edn. Berne, 1954), 148–54.

sequence of epistles leads readers along another series of *ambages*. While the absence of the Ardillon exchange in the Paris editions does not suppress any major themes absent from other liminary pieces, it does entail an absence of Latin prose, reducing the editions' formal diversity and thereby attenuating their labyrinthicity.[45]

This Paris–Poitiers contrast is reinforced by differences in the respective editions' layout. In the Poitiers editions, as previously mentioned, the marginalia use a different type from the French verse. Larger type is employed on the title-page and in certain headings; Ardillon's letter is in roman; and Bouchet's letters to Ardillon and Prévost use smaller bastard characters. Parisian printers vary their typefaces only on title-pages and in some headings; consequently, the *Labirynth*'s variety of languages, voices, and textual functions (narrative, annotation, liminary pieces, etc.) is not so strongly highlighted as in the Poitiers editions.

The differences in layout, illustration, and paratext reveal that the Paris editions place less stress than the Poitiers imprints upon Bouchet's self-conscious techniques, especially the labyrinth's implications. Yet this contrast is only relative. All editions include, first, copious liminary material, which underlines the importance of Bouchet's didactic metaphor and the prominence of his creative and compilatory activity, and second, dense annotation which demonstrates Bouchet's authority, and highlights and complicates the labyrinth image. Likewise each edition's title-page reproduces the same formula, which refers to Bouchet by his usual *Traverseur* pseudonym and identifies him as the *Regnars*' author (for the first edition's title-page, see Plate 5.8). This was perhaps originally a commercial strategy by Bouchet, both exploiting the publicity value of a striking appellation and reclaiming his authorship of the *Regnars* from its fraudulent publishers; however, it also presents Bouchet as primarily a poet, whose identity is defined by his previous works. His writerly activity is highlighted: the pseudonym appeals to readers' literary experience, besides aptly suggesting that Bouchet travels the *voies perilleuses* of a labyrinth. Moreover, the reference to the *Regnars* ties the *Labirynth* to Bouchet's past writing, suggesting he produces a coherent *œuvre*. Bouchet's later publications often refer to his

[45] Also absent in Paris editions is the *huitain* on the title-pages of Poitiers editions (see Plate 5.8).

LE Labirynth de fortune et Seiour des trois nobles dames Cōpo-se par lacteur des Renars trauersans/ et loups rauissans surnomme le trauerseur des voyes perilleuses.

Ex Lib. S[tae] Genovefae paris

Aux lecteurs
¶ Homme mortel qui ne veulx mescongnoistre
Ton dieu/ne toy/ce petit euure litz
Et tu verras comme on se peult congnoistre/
Et que souuent le mal pour bien eslitz/
Si tu le croys chasseras tous delictz
Dieu ameras sur tout et puis ton proche/
Regardes tout (combien quil soit prolix)
Auant quen dire aulcun mal ne reproche.

Cum Priuilegio.

¶ Et sont a vendre a Paris en la rue sainct Jacques deuāt sainct yues/ a poictiers deuāt le pallays au pellican par Enguilbert de marnef Et a limprimerie a la celle/ deuāt les cordeliers par Jacques Bouchet Imprimeur.

PLATE 5.8 Title-page of the first Poitiers *Labirynth*: Poitiers, Médiathèque François-Mitterrand, *Le Labirynth de Fortune*, DP. 119

earlier material in this way: two of the *Epistres morales et familieres*, for instance, list his works to date.[46]

Despite the importance of the *Labirynth*'s self-conscious aspects, its didactic value cannot be overlooked. The annotations' main role is doctrinal, and Bouchet's letter to Prévost presents piety and pedagogy as his prime concerns. Yet Bouchet's choice of narrative framework renders the poem's self-consciousness inseparable from its instructive content. Especially in the Poitiers editions, the *Labirynth*'s physical presentation develops Bouchet's sophisticated use of the labyrinth image, and his suggestive presentation of himself and his writerly practice.

Les Anciennes et modernes genealogies des roys de France

Bouchet's *Genealogies* is a piece of nationalistic French historiography, comprising a prose account of Frankish kings before Pharamond, and for each subsequent king a prose summary of his reign; a woodcut portrait; and a verse epitaph in the king's own voice. It was hugely popular: first printed in 1528, it was frequently reprinted in Poitiers and Paris, and in 1545 appeared in a luxurious folio volume along with revised versions of several other Bouchet pieces.[47] Its specially produced woodcuts, which probably contributed to the text's success, were even more influential: they were copied, with the verse epitaphs and a précis of some of Bouchet's prose, in Barthélemy de Chasseneuz's *Epitaphes des roys de France*.[48]

The *Genealogies* primarily serves to glorify France's monarchy; Bouchet adopts the usual prestigious myth of Trojan descent, and his accounts of exemplary and some not-so-exemplary rulers have

[46] Bouchet, *Epistres*, fos. h5^{v}, F2^{v}.

[47] The editions of which copies or facsimiles have been located are: Poitiers, Jacques Bouchet, 26 January 1527/8 (MFM, CR. 152); Poitiers, Jacques Bouchet, 27 November 1529 (MFM, DP. 1781); Paris, no pub., 1530 (Tchemerzine, *Bibliographie*, ii. 51–2); Poitiers, Jacques Bouchet, 27 November 1531 (BL, 596. b. 11); Poitiers, Jacques Bouchet, 12 June 1535 (MFM, DM. 843 (1)); Paris, Galliot Du Pré, 1536 (MFM, DM. 844); Poitiers, Jacques Bouchet, 16 July 1537 (MFM, DP. 116); Paris, various booksellers, 1537 (MFM, DM. 846); Paris, various booksellers, 1539 (Bodleian, 8° S. 231 Art); Paris, various booksellers, 1541 (BNF, Rés. L^{37}. 2. E); *Les Genealogies, effigies et epitaphes des roys de France*, Poitiers, Jacques Bouchet and Jean and Enguilbert de Marnef, 1545 (Bodleian, Buchanan c. 19). See also Britnell, *Jean Bouchet*, 318–20. I quote from the first edition throughout, unless otherwise stated.

[48] Britnell, *Jean Bouchet*, 116.

implicit didactic value. Since neither the current king, Francis I, nor his predecessor Louis XII, had been the previous king's son, it is significant that Bouchet strives to present the royal dynasty as an unbroken lineage. He frequently asserts continuity of royal descent, and therefore Louis's and Francis's legitimacy; he claims that the dynasty's traditionally accepted division into three 'diuerses generacions' only applies if one considers the male line of descent alone, and that the lineage is unbroken 'a cause des fẽmes voire des hõmes a bien le prendre' (fo. L6^{v}). The woodcuts' effect in glorifying the monarchy is especially important. They have been observed to form part of an increasing contemporary vogue for visually representing the royal lineage, also manifested in royal entries and the statues in the Grand' Salle of the Palais de Justice.[49]

While dynastic glorification is paramount, however, various aspects of the *Genealogies* draw attention to Bouchet's writerly investment in his text. As usual in his historical works, Bouchet stresses his scholarly efforts, especially his attempts to harmonize conflicting accounts in his sources: 'ie me euertueray dieu aydant de mõstrer en la deduction de ce petit euure sans aultre epilogacion que es oppiniõs cy dessus recitees ny a si grant contrariete quon pourroit de prime face iuger' (fo. B1^{v}). Occasionally he even compares his work advantageously with that of other scholars, thereby proving his authority. Robert Gaguin, for instance, is wrong in claiming there was a dispute between Henri I and his mother, 'mais ie nen blame ledit Gaguin leq̃l en toute sa cronique a tousiours suyuy vng liure barbare et tout corrumpu intitulle les grans croniques de france/ et na este curieux de senquerir de la verite auec les antiques historiographes esquelz foy doit estre adiouxtee' (fo. P3^{r}).[50] Elsewhere, Bouchet guides readers through his account, explaining its parameters and alluding to his previous work:

Iauois ꝓmis voire fait plus de la moictie des annalles de frace [*sic*]/ ꝑ lesq̃lles ieꝑsois escripre le sõmaire des faictz ⁊ gestes des roys de france et les cõcordances des croniques/ q̃3t aux generalogies [*sic*] dates ⁊ aultres discrepances Mais cõsiderant que la cronicq̃ de maistre Robert Gagin [*sic*] historiographe de grant

[49] Britnell, *Jean Bouchet*, 11b.

[50] For historians' criticism of Gaguin's excessive reliance on the *Grandes Chroniques*, see F. Collard, 'Une œuvre historiographique du règne de Charles VIII et sa réception: Le *Compendium de origine et gestis Francorum* de Robert Gaguin', 83–4.

renõmee est assez ample/ et que pour le peuple vulgaire elle a este traduicte τ mise en langue francoise/ et que par ce seroit vne redicte a moy ie me suis deporte de escrire en ce petit euure desditz faictz et gestes. Desquelz et de ladicte cõcordãce iay sõmairent [*sic*] ꝑle en mes annalles de guienne/[51] ou iay cõpris en brief le total des Croniques de france τ la cõcordãce dicelles. Et presentemẽt escriray seulement les genealogies/ τ epitahes [*sic*] des Roys. (fos. H2^{v}–3^{r})

Here, as with his use of the Traverseur pseudonym, Bouchet assumes the role of global author, responsible for an *œuvre* extending beyond the text currently being read.

The *Genealogies*' variety of forms and media complements this self-conscious vein. Prose, verse, and woodcuts together create the impression that Bouchet is embracing the complete possible range of linguistic and visual resources so as adequately to praise France's kings. This variety, however, is not accompanied by sophistication: prose and verse are stylistically plain. Rather than the richness of individual elements, it is their combination which provides a suitable sense of elevation. Nevertheless, the diverse linguistic and visual registers underline that the text has been artistically fashioned, thereby reminding readers of Bouchet's compositional activity.

To examine the relationship between the *Genealogies*' physical presentation and the literary self-consciousness it displays, I shall discuss first the *editio princeps* published in Poitiers, and then subsequent Poitiers editions. I shall proceed to contrast these texts with the editions printed in Paris, without Bouchet's participation, and finally to analyse the 1545 anthology. This approach permits assessment of the ways in which Bouchet's attitude towards the text's material form developed, and in which the *Genealogies* is presented according to different priorities in the absence of authorial influence.

The first edition prefaces the *Genealogies* with two Latin poems by Nicolas Petit and three texts by Bouchet, a verse epistle to the diplomat Marc de La Mothe au Groing, and letters to Antoine Duprat, Chancellor of France, and to François the Dauphin. Most of these pieces underline the *Genealogies*' patriotic flavour: the letter to the Dauphin, especially, stresses the Gauls' and Franks' ancient military prowess, and notes France's traditional upholding of the Church (fos. A1^{r}–7^{v}). Simultaneously, however, Bouchet's practice

[51] i.e. *Annales d'Aquitaine*; see Britnell, *Jean Bouchet*, ch. 4.

as scholar and writer is highlighted. Petit, whose verse in fact provides the only identification of Bouchet on the title-page (Plate 5.9), praises him for tracing the Frankish monarchy back beyond Pharamond, and for his brevity and accuracy; as in many other Bouchet works, material by other writers helps guarantee Bouchet's competence. Bouchet himself also draws attention to his activity in the La Mothe epistle:

> Iay concuilly de liures autenticques
> Long tẽps absconds en frãce/ τ tresanticques
> Lantiquite de tous ces nobles roys/
> Et concorde les discors et derrois
> De leurs parens et genealogies
> Depuis Francus τ mis les effigies
> Et nobles faictz des roys mors et enuers
> Puis Pharamond en vernaculles vers/
> Et le tout prins des historiographes/
> Puis redige soubz forme de Epitaphes. (fo. +5^{r})

The Duprat letter is especially significant, as Bouchet here defends his writing. He prefers to write and risk criticism than not write at all, in similar fashion to Diogenes,

> lequel non obstãt la derision de ceulx de Corinthe trouuoit meilleur transporter de lieu en aultre le tonneau ou il couchoit que estre oiseux τ sans rien faire/ non sans cause appelle ce que iescriz vng tonneau a mectre vin/ parce que mon intention est ne mectre en mes petiz escriptz autre chose que ne soit au vin ressamblant/ qui a couleur/ goust/ et force. (fo. +3^{r})

The comparison helps underpin Bouchet's authority by invoking ancient wisdom as a suitable frame of reference within which to assess his practice. Earlier in the letter, Bouchet also underlines the woodcuts' importance, by claiming his text is 'vng art de memoire' (ibid.). Ordered mnemonic images were fundamental to medieval and Renaissance memory systems, based upon ancient techniques; by drawing upon this tradition, Bouchet suggests that the *Genealogies* has a double memorial value, both commemorating individual kings and enabling the royal succession to be memorized.[52] He shows that the cuts are inseparable from his project, reinforcing the impression given by the book's privilege, which

[52] See Cowling, *Building*, 133, and, more generally, F. A. Yates, *The Art of Memory* (London, 1966).

Les Anciennes et modernes ge-
nealogies des Roys de France
et mesment du roy Pharamond
auec leurs Epitaphes et Effi-
gies.·.

¶Nicolai Parui Bellosanensis ad Ioãnem Bouchetum
vetustissime Francorũ historiæ parentem Epigramma.
¶Ignorabantur nostre primordia gentis,
Francorum primus, rex Pharamundus erat.
Quadraginta illum sed præcessisse recenses
O Bouchete grauis conditor historiæ.
Et q ab Hectoreo Francorum sanguine reges
Descendunt, etiam qui modo sceptra tenet.
Hoc tibi debemus studio qui prisca fideli
Gallorum primus nominis ista doces.

¶Et sont a vendre a Paris en la rue sainct Jacques
Et a Poictiers au Pellican. Et a limprimerie a la
Celle/ et deuant les Cordeliers par Jacques Bouchet
Imprimeur audict Poictiers.

¶Cum priuilegio.

PLATE 5.9 Title-page of the first *Anciennes et modernes genealogies*: Poitiers, Médiathèque François-Mitterrand, *Les Anciennes et modernes genealogies*, CR. 152

mentions Bouchet's wish to illustrate his text with 'plusie^rs belles hystoires mesmemẽt desd' effigies desd' Roys'.[53]

Indeed, the woodcuts play a very specific role in the text. Besides the portraits of kings, the first edition bears the French royal arms on its title-page, and a cut depicting Francus, Aeneas, and Antenor leaving the burning city of Troy after the liminary pieces (fo. A8^r). The former image underlines the *Genealogies*' nationalistic flavour, especially since the arms are accompanied by the imperial crown: this reflects Bouchet's assertion that the Holy Roman Empire has only passed to the Germans because they are 'descẽduz des frãcois a cause des fẽmes' (fo. N6^r). The latter illustration, first used in Bouchet's *Histoire et cronicque de Clotaire*,[54] highlights the French claim to Trojan descent. Moreover, for readers familiar with Bouchet's output, the cut evokes the *Histoire*, once again presenting Bouchet as the author of a wide *œuvre*.

All the royal portraits except the first are numbered; this both aids compositors and suggests to readers a balance between dynastic continuity (a single numerical series) and the individuality of each king. They are set in ostentatious frames, of which there are three designs, dominated by motifs such as putti, masks, vine-scrolls, candelabra, pendants, and tablets (e.g. Plate 5.10). Hence the frames have a classicizing flavour, in counterpoint to the bastard typeface of the text, whose connotations are strongly vernacular. This dual visual discourse reflects both the *Genealogies*' nationalistic orientation, through the typeface, and the antiquity of France's royal lineage, through the frames. The images of the kings themselves are all unique; differences between them, however, are often minor, and usually achieved by varying kings' headgear or accompanying attributes, the angles of their faces, or the length of their hair or beards. Actual resemblance, other than for recent rulers like Charles VIII, whose famously hooked nose is depicted (Plate 5.10),[55] is a low priority: Charles the Bald has hair and a prominent beard. It suffices that each king looks different; combinations of distinctive features enable kings to be differentiated and thus memorized effectively.

[53] The privilege, dated 24 April 1528 and granted to Enguilbert de Marnef for four years, does not appear in all copies of the *editio princeps*, whose publication it post-dates (Davies, *Catalogue*, i. 52–4). I cite from the second edition, fo. +1^v.

[54] Poitiers, Enguilbert de Marnef, n.d. [1518] (BNF, Rés. 4° L^4 b. 4), fo. A1^r. Britnell, *Jean Bouchet*, ch. 4, discusses this text.

[55] See Scheller, 'Imperial Themes', 29.

Les Epitaphes
en labbaye saīct denys en frāce. Des le viuāt
dud' roy Loys son pere il pmist prēdre a fēme
Margarite de flādres fille de lēpereur Maxi-
miliā laq̄lle po^z ceste cause fut amenee en frā
ce Et apēs q̄ led roy Charles eut eu la victoire
cōtre frācois duc de Bretaigne ⁊ q̄ iceluy fran
cois fut decede il espousa sa fille aisnee mada
me anne ⁊ rēdit lad margarite aud maximiliā
auec la cōte dartoys dōt il se cōtēta/il eut plu
sieurs enfans de lad Anne mais to⁹ moururēt
parquoy faillit en luy la ligne directe de ceulx
de Valoys. ¶ Sensuit son Epitaphe.

PLATE 5.10 The unmistakable Charles VIII in the first *Genealogies*: Poitiers, Médiathèque François-Mitterrand, *Les Anciennes et modernes genealogies*, CR. 152, fo. V8^v

The cuts' positioning, between the prose narratives and verse epitaphs, affects their function significantly. The title *epitaphes* suggests the verse pieces are tombstone inscriptions: contemporary poems entitled *épitaphe*, especially those in the first person such as Bouchet's pieces, often had this quality.[56] Indeed, some of Bouchet's epitaphs include references to 'ce marbre' or 'ceste sepulture' (fos. N8^{v}, Q1^{v}), implying they are genuine inscriptions. In this context, the woodcuts inevitably become more than mnemonic images, and attain the quality of tombstone effigies; Bouchet's description of the cuts as *effigies* is highly appropriate. Moreover, the sepulchral quality of *effigies* and *epitaphes* in combination implies that the book is a metaphorical tomb for France's monarchy. While Bouchet's didactic works often contain descriptions of memorial buildings, here the monument is to be identified with the whole text, and indeed with the book as object: the explicitly inscriptional connotations of *effigies* and *epitaphes* underline the volume's materiality, the printer's activity which has produced the book. Thus the cuts not only exalt France's kings, but contribute to an equivalence of text and tomb. Readers are made aware that the text is a constructed artefact, a creation of both language and paper. Bouchet's writerly activity, and the material processes of textual production, show through the royal subject-matter.

The cuts' position also entails that they mark formal shifts, from prose to verse and from third- to first-person discourse; they may even be regarded as pictorially identifying the speaker of each epitaph. Their articulatory role is all the more useful because there is little spacing between sections of text. Besides the presence of images, and the different appearance on the page of blocks of prose and verse, most visual variety in the *Genealogies* occurs amid the liminary material. Here, the bastard typeface used for most of the text is often replaced: Petit's verse is in roman, appropriately for a Latin text, and the privilege and Duprat and La Mothe epistles in a smaller bastard. Type also varies in size in a few headings, while small type is employed for occasional marginal annotations. These are not citations or source references, but summaries of the episodes currently being narrated, which assist browsers. With the illustration and variation of type, they vary the reader's visual experience.

[56] Molinet's *Epitaphe du duc Philippe de Bourgogne* (*FD*, i. 34–5) is an influential example.

The *editio princeps*, then, displays characteristic features of Bouchet editions: liminary material thematizes Bouchet's practice as writer, and guarantees his authority. Less typically, the woodcuts not only contribute to the *Genealogies*' formal variety, but also underpin the text–tomb equation which further highlights Bouchet's role in recounting and commemorating the deeds of French kings.

Subsequent Poitiers editions retain the first edition's quarto format, excepting that of 1537, an octavo volume. Typographic variation gradually diminishes: the second edition abandons smaller type for the Duprat and La Mothe epistles, while in that of 1535 the only typefaces other than the standard bastard are roman type for the Latin verse on the title-page and occasional larger typefaces in headings. This same edition also lacks the privilege, Petit's second poem, and the marginal summaries; these do not reappear in the 1537 edition. All editions from 1529, however, underline the epitaphs' physical, inscriptional value by introducing them with decorative woodcut initials, in contrast to the prose summaries which begin with paraphs. All Poitiers editions use the same woodcuts and general layout, with a few exceptions.[57] The 1537 edition's limited page size precludes the use of frames round the images, but these appear in all other editions. In general, the successive Poitiers editions evince a slight overall decrease in visual variety, though without radically altering the first edition's main effects. Whether Bouchet sanctioned the changes is uncertain, but it is surely significant that Poitiers editions depart from the *editio princeps* far less than those produced in Paris.

The earliest Paris edition, published in 1530 in contravention of Enguilbert de Marnef's privilege, immediately departs from the Poitiers tradition by adopting an octavo format. This becomes the standard format for Paris editions; Galliot Du Pré's 1536 edition, a sextodecimo, is the sole exception. Probably due to its exiguous format, this latter edition has no woodcuts, a surprising feature since Du Pré specialized in illustrated books.[58] As a result, the *Genealogies*' form and function are fundamentally altered. Textual

[57] The second edition replaces the original cut of Charles VI, possibly because this had depicted him with a beard, contravening the principle adopted under Louis VII whereby 'les Roys de frãce τ aultres cesserẽt de porter longues barbes' (*editio princeps*, fo. P8v). The original Charles VI cut reappears in the 1535 and 1537 editions, replacing the earlier cut of Theodoric II; it also depicts Clotaire IV in 1535, and Philip the Fair in 1537.

[58] A. Charon-Parent, 'Aspects de la politique éditoriale de Galliot Du Pré', 209.

references to *effigies* must be suppressed: consequently, the Duprat letter is omitted entirely, and along with it Bouchet's striking defence of his writing. Petit's Latin material is also absent; Du Pré innovates by employing roman type, but type size varies only in headings. Other Paris editions depart less radically from Poitiers precedents, but nevertheless display important differences. The collectively published editions of 1537, 1539, and 1541 use a set of cuts first employed in 1530, much smaller than the Poitiers cuts, of which they are mostly crude copies. These images lack frames, and appear at the top left of each prose summary, rather than between the prose and verse. Consequently, they lose some of their value as effigies by being dissociated from the verse, and now mark different reigns rather than formal divisions. Their memorial function is maximized, at the expense of their role in foregrounding the *Genealogies*' self-conscious aspects. In the 1537 edition, the cuts do not even distinguish the section of summaries and epitaphs from the previous historical account, because woodcuts also depict most of the Frankish kings before Pharamond. Portraits, then, are no longer peculiar to Pharamond's lineage. Several of these cuts remain in the 1539 and 1541 editions, though most are omitted. Paris editions also differ from Poitiers imprints by lacking marginal summaries and typographic variation, and by including less liminary material: the 1537–41 editions use bastard type, with larger type only in a few headings, and lack Petit's second poem. Furthermore, the distribution of woodcut initials no longer suggests the epitaphs' inscriptional quality in contrast to the prose summaries. Both prose and verse sections have initials in the Paris editions, so that the verse appears no more material than the prose. The 1541 edition reverses the Poitiers editions' tendency, by employing initials for most prose sections but merely *lettres de tourneure* for the verse.

Thus the Paris editions' presentation blocks, to an extent, the text–tomb equivalence which Poitiers editions highlight. Formal variety is attenuated; the *effigies* and *epitaphes* become very much secondary to the historical account. While the frequent reprintings and small formats imply that Bouchet's text enjoyed a wide readership of relatively modest means, Paris readers had a much poorer visual experience than their Poitiers counterparts, and received a narrower interpretation of the *Genealogies*' function.

The 1545 anthology printed in Poitiers, including the *Genealogies* under a revised title, marks an important phase in the textual history of Bouchet's works. In this year, as previously mentioned, Bouchet undertook a systematic project of producing earlier works in definitive and sumptuous folio editions. Under a privilege granted him in January 1544, he published the *Epistres morales et familieres*, a new work, though the epistles were mostly of earlier date;[59] the fourth of five authoritative updatings of his historical work the *Annales d'Aquitaine*; and the *Genealogies, effigies et epitaphes*. Here, in a new liminary piece entitled 'Au Lecteur', Bouchet explains his undertaking:

> I'ay reueu, corrigé & amende [*sic*]. scelon les modernes Orateurs, aulcuns de mes opuscules, & iceulx remis en leur ordre au plus pres qu'il m'a esté possible du style que les Orateurs de ce regne gardẽt & obseurẽt [*sic*]: lesquelz i'ay faict Imprimer, & aultres que ie n'auois mis en lumiere, en cinq Tomes, dont cestuy est le troisiesme,[60] auec Priuilege de quatre ans, par l'octroy & congé du Roy nostre souuerain seigneur . . . Vous suppliant benings Lecteurs aulcunement m'excuser si aulcunes faultes trouuez en ce que i'ay escript, ayant cõsideration a mon eage septuagenaire,[61] & a l'estat de practicque Palatine que i'ay principallement exercé. (fo. aa6^{v}; all subsequent references are to this edition)

Bouchet here expresses explicitly many tendencies common in his work from the 1520s onwards. His reworkings of older texts both reinforce the importance of form, since they often entail alterations in accordance with recently formulated prosodic conventions, and evoke an image of Bouchet as an author whose identity embraces (and is defined by) his whole *œuvre*. His diverse works cohere into a global project, over which he ostensibly has full control. While the apparently autobiographical references to his age and job as *procureur* anchor Bouchet in the real, the thematization of his writerly practice underlines his compositional and scholarly investment in the *Genealogies* and other pieces.

Besides the new privilege and letter 'Au Lecteur', the 1545 collection includes the Duprat and Dauphin letters, but not Petit's verse or the La Mothe epistle. These alterations have little significant effect upon the *Genealogies*' content, since the major themes of the excised pieces are also present in those which

[59] Britnell, *Jean Bouchet*, app. B.

[60] The fourth and fifth volumes were apparently never printed.

[61] Born in January 1476, Bouchet was in fact not quite 70.

remain. Rather, they suggest a broadening of Bouchet's appeal. The use of individual luminaries as literary guarantors is attenuated in favour of a direct appeal to the general reader, reinforcing the impression of Bouchet's controlling presence. Bouchet adopts this strategy quite frequently in his later work. Besides general prefaces accompanying more personal liminary epistles in the *Triumphes* and his commemoration of Francis I, the *Triomphes du tres chrestien roy de France*,[62] an address 'Au Lecteur' appears in his manual of legal procedure, the *Forme et ordre de plaidoirie*,[63] and a piece declaring 'Le motif & intention du Trauerseur' in the *Epistres* (fo. $\pi 6^{r}$). Bouchet acknowledges and exploits the way in which printing enables texts to reach a wide, anonymous audience.

The *Genealogies* proper employs the same woodcuts as the previous Poitiers edition, from 1537, positioned according to standard practice in the Poitiers editions. However, the folio format entails greater visual diversity than the previous quarto and octavo volumes had permitted. While prose stretches across the page in wide blocks, verse is left-justified in a single column, with copious white space to the right. Prose and verse are thus maximally distinguished, and an impression of luxury given which reinforces Bouchet's glorification of the French monarchy: a more economical two-column verse layout is possible, but has not been adopted, as if wastage of paper were no object. The text's visual richness is accentuated by variation of typeface: headings use a smaller roman type than the body of the text.

Most importantly, this new edition contains several other texts anthologized after the *Genealogies*. These comprise a *Deploration de feu Françoys de Valoys, Daulphin de Viennoys*, published for the first time; a series of *Epitaphes de plusieurs personnes*, many previously unpublished;[64] the first edition of a *Deploration de feu monseigneur, monsieur Françoys de La Tremoille*; new versions of the *Chapelet des princes*, *Rondeaux et ballades*, and *Déploration de l'Église militante*, which all appeared in Bouchet's *Opuscules* collection of 1525 as well as earlier, separate versions;[65] newly published pieces entitled *Dizains moraux sur les apophtegmes et subtiles responses de sept sages de Grece* and

[62] Poitiers, Jean and Enguilbert de Marnef, 17 August 1549 (BNF, Rés. Ye 191), fos. & 2^{r-v}, & 5^{r-v}.

[63] Poitiers, Jean and Enguilbert de Marnef, 1542 (MFM, DM. 1402 (1)), fo. ã2^{r}.

[64] See Britnell, *Jean Bouchet*, 313.

[65] See ibid. 303, 306, 309, 316.

Aultres Dizains de plusieurs matieres; the *Angoysses et remedes d'amours*, previously published separately;[66] and new groups of short verse pieces, *Quatrains et cinquains d'aulcuns memorables faictz* and *Patron pour les filles qui veulent apprendre à escripre*. The texts remain relatively distinct, many beginning on a new page; the *Angoysses* even has its own title-page (fo. A5^r^). While the collected material is thematically quite coherent, dominated by epitaphs, laments, and didactic verse, there is considerable formal variety, reflecting that within the *Genealogies* proper. Verse forms vary greatly: the *Déploration* consists of *treizains*, the *Chapelet* of ballades and rondeaux, and the *Angoysses* of differing forms, including decasyllabic couplets, octosyllabic *huitains*, and heterometric quatrains. Linguistic diversity is also apparent: the *Dizains moraux* expound Latin maxims, printed before each *dizain*. The use of smaller type for Latin visually reinforces this variety, as does variation in the use of columns. All prose, the *Genealogies* verse, and the material from the *Opuscules* use a single column; other verse adopts two.

Some of the anthologized pieces draw attention to Bouchet's compositional artistry. The *Chapelet*, for instance, which offers moral guidance to princes, is intricately structured. A heading informs its readers that its fifty rondeaux and five ballades (plus a prefatory ballade) are arranged so that 'Apres le dixiesme Rondeau trouuerez tousiours ballade' (fo. q1^r^). This organization, which echoes the construction of a coronet or rosary, suggests a piety appropriate to Bouchet's didactic project, and clearly indicates his craftsmanship.[67]

Bouchet's investment in his writing is foregrounded rather differently in the dedication which begins the *Angoysses*, addressed to Louis d'Estissac. A *remedium amoris*, in which a first-person narrator overhears four unhappy lovers' laments and a diatribe against love from Pallas Athena, the *Angoysses* had a lengthy prehistory. It was first published in Paris early in the sixteenth century, in a pirated version entitled *L'Amoureux transi sans espoir*;[68] not until 1537 did an authorized edition appear. In the dedication, Bouchet recounts the piracy, and outlines the changes he has since made to the text's versification:

[66] See ibid. 325. [67] Ibid. 91–5 analyses the *Chapelet*.

[68] See M. B. Winn, 'Publisher vs. Author: Anthoine Vérard, Jean Bouchet, and *L'Amoureux transy*'; *Anthoine Vérard*, 81–3.

aulcuns Imprimeurs de Paris, ou lors faisois demourance, plus desireux du remplissement de leurs bourses, que de leur honneur, ne du mien, auoient trouué moyen de retirer partie de mes compositions petites, & les auoient incorrectement imprimees, & a icelles baillé nom & tiltre a leur plaisir . . . Ie ne synalymphois lors les quadratures de la rithme de dix & vnze piedz, comme ont tousiours faict Georges, Clopinel, Castel, Iehan le Maire, & aultres irreprehẽsibles orateurs Belgiques, qui est necessairement requis. Et en rithme plate, qu'on appelle Leonyne, ne ordonnois ne entrelassois les masculins & femenins vers. (fo. A5^{v})

As in the preface to the revised *Regnars*, Bouchet narrates his work's genesis. In addition, he stresses his attention to prosody, setting himself in the context of other *rhétoriqueurs* who constitute yardsticks by which to judge his technical skill. He appeals not only to readers' moral sensibilities, through 'la reprehension des folz amoureux' (ibid.), but also to their literary connoisseurship.

Besides such indications of Bouchet's practice, the *Genealogies* volume also contains apparently autobiographical details. In an epistle to Lazare de Baïf, Bouchet complains about being taxed for five *écus*:

> Prester! qui? moy! chargé de huyct enfans,
> Et qui ay pres de soixante neuf ans? (fo. A4^{v})

Such laments admittedly have a long literary pedigree, and need not refer to genuine circumstances.[69] Nevertheless, the frequency with which Bouchet writes personal pieces—one need only think of the *Epistres* collection–implies there may well be some truth in his complaint. More importantly, as in other Bouchet works, readers are encouraged to identify the *je* on the page with the historical author Bouchet; the author's presence in his work is thereby accentuated.

By virtue of such features, the pieces accompanying the *Genealogies* proper in this 1545 edition increase the incidence of self-consciousness in the volume. They thus provide a context within which readers are more likely to perceive the *Genealogies*' self-conscious elements, rather than regarding it solely as dynastic propaganda. In this sense, the 1545 edition marks the culmination of Poitiers publishing practices, which tend to accentuate the *Genealogies*' formal heterogeneity and the text–tomb equivalence. Bouchet's continual involvement in publication is surely a major

[69] See C. Scollen-Jimack, 'Marot and Deschamps: The Rhetoric of Misfortune'.

influence upon this trend. In contrast, Paris editions, over which Bouchet presumably had no control (witness the 1530 edition's illegal publication), present the *Genealogies* as primarily a historical work.

Conclusions

The texts analysed exemplify the tendencies in the presentation of Bouchet editions. Principal among these are: the use of liminary material and annotations to assert Bouchet's controlling presence and secure his authority; the exploitation of paratext to reinforce the impression which texts give of belonging to a wider, coherent *œuvre*; and the constant revision and updating of material, through which the image of Bouchet communicated to the reader is also redefined. Significantly, editions produced without Bouchet's involvement usually attenuate the extent to which physical presentation highlights texts' self-conscious aspects; hence the frequent contrasts between editions printed in Poitiers and those published elsewhere. Though particularly clear in the *Genealogies*, this tendency also appears in other works. One such example is provided by the *Cantiques et oraisons contemplatives de l'ame penitente*, a collection of vernacular prayers in verse probably composed before 1520. These first appeared in BNF, n. a. fr. 11555, a well-produced and clearly presented manuscript, containing long explanatory rubrics (distinguished from the prayers by red ink) which underline Bouchet's pedagogical project.[70] In a dedication to Hélène de Hangest (fo. 1^{r-v}), Bouchet defends his use of French as a means of enabling uneducated people to understand their prayers.[71] Part of the collection was later printed; the surviving copy cannot be dated or localized, though it is bound with a copy of Poitiers Hours printed in Poitiers in 1525.[72] The printed version lacks the dedication, and is laid out less clearly: a single bastard typeface is used throughout, while white space is rarely employed to separate headings from the texts they introduce, and never between stanzas. Though textual alterations can be attributed

[70] See J. Britnell, 'Jean Bouchet's Prayers in French for the Laity, *Les Cantiques et oraisons contemplatives de l'âme pénitente traversant les voies périlleuses*'.

[71] Edited ibid. 426.

[72] *Heures à l'usaige de Poictiers*, Poitiers, Jehan Coussot for Jacques Bouchet and Nicolas Pelletier, 20 December 1525 (BNF, Vélins 1655). See Britnell, *Jean Bouchet*, 312.

to new religious circumstances,[73] the presentational differences suggest that Bouchet's influence over textual production was less strong than in the case of the manuscript.

As the works discussed illustrate, Bouchet's editions frequently exploit layout and illustration.[74] However, they resemble Molinet and Lemaire editions in that their physical presentation rarely foregrounds texts' formal virtuosity: the acrostic in the *Regnars* is a relatively uncommon exception. Yet the *Regnars*' and *Genealogies*' intricate text–image interaction, and the *Labirynth*'s use of dense marginalia to extend and thematize the labyrinth image, show that Bouchet's editions usually draw attention to language and imagery more effectively than do editions of Molinet's or Lemaire's poetry. The lack of stress on technical skill is primarily due not to the editions' presentation, but to Bouchet's own lack of interest in such considerations: those constraints he does set himself, such as the alternation of masculine and feminine rhymes, do not immediately strike readers as *tours de force* in the same way as Molinet's prosodic acrobatics.

Indeed, the notable aspects of Bouchet's career—his lack of a court position, the dominance of print in the transmission of his works, the insistent authorial presence in text and paratext, and the lack of ostentatious artistry—seem interrelated. Lacking a position at court, Bouchet relied more than most *rhétoriqueurs* upon print; printing, in turn, enabled him to authenticate and nuance his arguments, to draw attention to significant features of his work, and to ensure the prominence of his name or pseudonym. To assert himself as a poet, then, Bouchet could highlight his authority and compositional practice more directly than earlier poets, who depended more closely upon patrons, and had less control over their texts' transmission. In this respect, despite his relative conservatism as a poet, Bouchet anticipates the more apparently personal and less overtly formalist poetics of the mid- and late sixteenth century.

[73] See Britnell, 'Jean Bouchet's Prayers', 435–6.

[74] Other significant illustrated editions are those of *L'Histoire et cronicque de Clotaire*, with eight cuts, and *Le Jugement poetic de l'honneur femenin* (Poitiers, Jean and Enguilbert de Marnef, 1 April 1538; MFM, DP. 113) with eleven. The 1541 Rennes edition of the *Triumphes de la noble et amoureuse dame* contains numerous illustrations, hand-coloured in the sole surviving copy (BNF, Vélins 586): see Britnell, *Jean Bouchet*, 323; A. de La Borderie, *Archives du bibliophile breton* (Rennes, 1880–1907), ii. 91–100.

CONCLUSION

The preceding chapters trace an evolution in *rhétoriqueur* poetics, under the influence of printing and the new conditions of textual transmission and presentation it fostered. Molinet's poetry, produced in a manuscript-dominated culture, is characterized by complex language and versification, which are frequently highlighted in manuscripts. While his work appears in a wide range of anthologies, with differing priorities, formal features are usually given greater prominence than historical or didactic content. Less consistent is the extent to which his poems are attributed in manuscripts; while some anthologies are devoted to Molinet's work alone, others (especially earlier volumes) reproduce his pieces anonymously. The formal self-consciousness of Molinet's poems is acknowledged much less often in print. In the *Faictz et dictz*, technical virtuosity is secondary to didactic and political discourse, while editions of the *Temple de Mars* typify single-text editions in privileging the poem's perceived relevance to current affairs. The editions printed in Valenciennes, in whose publication Molinet probably played a part, are unusual in balancing political and formal elements. Attributions are rare outside Burgundian areas, with the important exception of the *Faictz et dictz*, an author-centred anthology. The differences between manuscript and printed versions of Molinet's poetry illustrate the way in which, in the absence of authorial influence, printing's technical and commercial imperatives encourage the publication of texts with a presentation which foregrounds their topical or morally practical value, rather than intricate prosodic devices which appeal to an audience of initiates.

The work of Jean Lemaire de Belges displays a wider range of self-conscious features than Molinet's. Allegory, versification, intertextual play, and metatextual discourse are all significant aspects of different texts. Manuscripts of Lemaire's poetry reflect this variety in their material presentation. While early editions similarly foreground diverse techniques, the emphasis increasingly falls upon the way in which Lemaire thematizes poetry and the activity of writing, and upon his life and interests outside the world of the text. This is particularly the case with the material accompanying the *Illustrations*, exemplified by editions of the *Épître du roi*; here, a carefully delineated political stance, authorial self-promotion, and metadiscursive reflection are paramount. The transmission of Lemaire's work thus continues the shift, identified in editions of Molinet's poetry, away from primarily formal and linguistic considerations. Yet in Lemaire's case, this is not solely a matter of editions failing to reflect texts' self-conscious features. Rather, after his earliest works, Lemaire focused less exclusively on language and versification, and more on other manifestations of his compositional investment. He was also more deeply involved than Molinet in his works' publication, itself a symptom of changes in the institution of literature. The development of Lemaire's literary priorities seems to reflect an awareness of print's implications. To an extent, he came to write with an eye to publication in the modern sense; whereas 'publication' for Molinet seems primarily to have consisted in presenting a newly written text to a limited court circle, for Lemaire it increasingly entailed having a text printed for wide distribution. Yet, as the *Épître* and *Concorde des deux langages* illustrate, differences in emphasis between manuscript and printed versions often remain, perhaps because the two media appealed to different audiences. Manuscripts still attracted a restricted readership of cognoscenti, while print reached a wider and less knowledgeable public.

Bouchet's poetry is predominantly transmitted in print, and Bouchet himself recognizes and assimilates the possibilities offered by print more fully than previous poets. His ongoing involvement in publication, and frequent reworking of his texts, reveals that publication was now a factor with which a writer had to reckon at all stages of his work. Bouchet solicits the larger, less well-read audience made accessible by print, by disseminating Latinate erudition in the vernacular, and by employing forms less elaborate than those of earlier *rhétoriqueurs*. Yet at the same time, editions of his work effectively convey his use of language and imagery, while paratext gives Bouchet a prominent, authoritative presence in his work, and helps subsume individual texts into a global and constantly evolving *œuvre*.

The development in the modes of literary self-consciousness displayed by these three poets can be illuminated further by recourse to the Barthesian distinction between the *texte lisible* and the *texte scriptible*, which I have previously evoked to help contrast the manuscripts and editions of Molinet's poetry. The *texte lisible* manifests a reassuringly clear relationship between signifier and signified, so that its language appears a transparent screen through which to view a real world; its readers have little interpretative liberty, and are cast as passive consumers. In the *texte scriptible*, by contrast, signifiers do not relate to signifieds in unquestioned fashion; readers must engage with the text's linguistic surface, instead of looking through it. Consequently, readers have an active role, working to produce the text.[1] Clearly, Barthes's terms cannot viably be considered as denoting the two sides of a dividing line, on one or the other side of which any text must fall. Rather, they represent theoretical extremes, the ends of a spectrum along which texts can be positioned. Any text has both *lisible* and *scriptible* elements: even radically experimental novels tend to display some mimetic features, while, as reader-response criticism has shown, readers can never be rendered completely passive.

If the works of Molinet, Lemaire, and Bouchet, and their manifestations in manuscript and print, are considered in terms of the *lisible* and the *scriptible*, a clear drift from *scriptible* to *lisible* may be identified in two areas. First, the physical presentation of texts attenuates *scriptible* elements. As printing supplants the manuscript, there is a normalization of the most *scriptible* texts, the multidirectional pieces and rebus-poems which require readers to construct the poetry for themselves. Technical and economic factors constrict spatial form, an inherently significant feature of verse, and one especially important in poems such as Molinet's *Ung present fait a l'empereur*, which is closer to the *lisible* end of the spectrum in the 1531 *Faictz et dictz* than in manuscript *A*. Secondly, new conditions of textual production affect techniques of composition. Bouchet's work is more straightforward than Molinet's: though Bouchet makes sophisticated use of allegory, he rarely employs the kind of intricate versification which arrests the attention of Molinet's readers at the level of sound and language. As his justifications of writing in the vernacular show, Bouchet makes virtues of clarity and intelligibility; his relatively unsophisticated prosody reflects these priorities. His practice seems to respond to

[1] See Barthes, *S/Z*, 11–16, 179–80, 205–6; T. Hawkes, *Structuralism and Semiotics* (London, 1977), 113–15.

the way in which printing required him to anticipate a readership less accustomed to complex forms than Molinet's public had been, by making texts accessible to a wider public which had more limited experience as readers of poetry than did the audience of manuscript-based poets. Lemaire's use of language, versification, and metadiscourse places him between Molinet's more *scriptible* work and Bouchet's more *lisible* writing, appropriately for a poet occupying an intermediate position between a manuscript-based and a print-based aesthetic.

The shift in *rhétoriqueur* poetry towards the *lisible* is accompanied by a more prominent authorial discourse, and by increasingly common attributions of texts to their authors. Manuscripts had already manifested instances of metadiscourse, and of a prominent 'author function'; the latter phenomenon, outlined by Michel Foucault, may be defined as the assignation of a work to a named author, the sense that a work's authorship is significant and relevant to our understanding of it, and the sense that the various works of a single author display some kind of coherence.[2] It was print, however, which helped make the author function a more pervasive form of literary self-consciousness than formal sophistication, at least in *rhétoriqueur* poetry.

This development is undoubtedly related to a phenomenon in early print culture which might be termed, in a *clinamen* from Harold Bloom, the 'anxiety of misreading'.[3] In several ways, writers manifest concern that their work may escape their control and be presented in a misleading, or even dangerous, light. Scholars quickly recognized printing's ability to disseminate mistakes,[4] while French writers were concerned by economics as well as textual accuracy; hence they sought privileges, brought lawsuits against printers, criticized unauthorized publications, addressed their own texts in worried tones, and attempted to guide reception through the paratext.[5] Such anxieties surface far more often than in manuscript culture, even though manuscriptural *mouvance* similarly threatens textual accuracy. The key difference is that

[2] M. Foucault, 'Qu'est-ce qu'un auteur?' Chartier, *Ordre*, 60–7, discusses manifestations of the author function in manuscripts.

[3] H. Bloom, *A Map of Misreading* (Oxford, 1975), 97, elucidates *clinamen*, a poet's swerving away from his precursor.

[4] Widmann, *Nutzen*.

[5] Besides the instances of such practices discussed above, numerous cases are analysed in Brown, *Poets, Patrons, and Printers*.

printed books could reach new, unknown readers, who lacked the familiarity with the author or his works which generally assisted the initial audience of a manuscript text (whether correct or not) in interpreting it. Patronage remained the standard form of employment for poets, yet it was increasingly uncommon for the court to constitute a poet's entire constituency. Even poets who resisted having their works printed, such as Guillaume Cretin, were published after their death, and sometimes before.[6] Strategies had to be developed to accommodate a wider circle of readers.

But such strategies also needed to bolster the author's more precarious position with regard to new conditions of book production. Sixteenth-century poets were confronted with a vast, impersonal apparatus, involving a heterogeneous printing workforce,[7] and commercial imperatives reflecting the supply of paper, the general level of consumer demand, and other elements of a global economy. The processes of manuscript production, less industrial in scale, were much easier than printing for an outsider to grasp. While the workforce in manuscript production might be divided into scribes, rubricators, illuminators, and binders, the division of labour in printing was far more complex. Because manuscripts were produced individually, their confection did not depend so urgently upon public demand as did the printing of a text in multiple copies; similarly, raw materials were required in much greater quantity by printers, so that their supply was more difficult to ensure. The increasing prominence of a named author, and of ostensibly personal discourse, are partly symptomatic of poets' attempts to assert their artistic personalities in the face of both a threateningly incommensurable publishing industry, and technical and economic factors which made formal virtuosity difficult to foreground in print. By the mid-sixteenth century, the author function is sufficiently significant to act as a point around which texts coalesce, whether for commercial reasons or on the basis of perceived similarities to named authors' works. Printed anthologies of the works of Alain Chartier, François Villon, Clément Marot, and, in England, Geoffrey Chaucer are not only author-centred; they also often include apocryphal material

[6] See Cretin, *Œuvres*, p. ci.

[7] See Febvre and Martin, *Apparition*, ch. 5; Parent, *Métiers*. Winn, *Anthoine Vérard*, ch. 3, discusses Vérard's relationships with various authors.

attributed to the poets in question.[8] The author's name exerts a kind of gravitational field, drawing other texts into its orbit.

Indeed, as printing develops, so the function of the poet's persona evolves. This evolution may be identified and described in simple terms, by regarding the first-person voice as a voice which may operate on two axes. It may function 'vertically', by apparently reflecting the historical author; in this case, it encourages readers to interpret the text biographically, to discern the historical poet's voice in the text. Alternatively, it may function 'horizontally', by resonating thematically and structurally with other elements of the text; it then assumes a signifying role immanent to the text, not a referential role with respect to the *hors-texte*. These two functions must be conceived, like the *lisible* and *scriptible*, as poles which can never be fully attained. The *je* cannot be wholly 'vertical', since the intertextual nature of all literature precludes the author's unmediated presence in the text. Neither can it be exclusively 'horizontal', for the very existence of a first-person voice raises the question of autobiographical reference.

In late-medieval French poetry, the first-person voice frequently admits of vertical elements, in contrast with the more horizontal *je* of the *grand chant courtois*. Various recent studies have penetratingly analysed this tendency in poetry from the fourteenth to the mid-fifteenth century.[9] Poetic personae of this period have multiple, shifting functions, though a broad shift towards increasing verticality is perceptible. Cynthia Brown, albeit without using the terms 'vertical' and 'horizontal', has identified occasional vertical features in the personae of the *rhétoriqueurs*' acknowledged forerunners, Jean de Meun and Chartier, and argued that printing intensified the development of a more vertical first-person voice. *Rhétoriqueur* poetry, according to Brown, evinces 'an evolution from a marginalized, narrative voice enmeshed in the fiction of the *récit* [i.e. a horizontal voice] to a more independent and aggressive first-person voice: that of the *acteur* who, standing apart from characters within the text, reacts instead with readers outside and identifies with the author himself [i.e. a largely vertical voice].'[10]

[8] See Chartier, *Poetical Works*, 143; Regalado, 'Gathering', 98; Marot, *Œuvres*, i. 9–11; A. Miskimin, *The Renaissance Chaucer* (New Haven, 1975), ch. 8.

[9] See Johnson, *Poets as Players*, chs. 2–4; de Looze, *Pseudo-Autobiography*, chs. 1, 3, 4; C. Attwood, *Dynamic Dichotomy: The Poetic 'I' in Fourteenth- and Fifteenth-Century French Poetry* (Amsterdam, 1998).

[10] Brown, *Poets, Patrons, and Printers*, 197.

My findings complement and confirm those of Brown, though situating this development within a shift in the self-conscious techniques which authors deployed, rather than primarily within a change in their socio-economic position. The authorial persona of Molinet, the earliest of the poets I have studied, operates both vertically and horizontally, though the latter function is paramount until late in Molinet's career. As I have previously indicated, the *je* in much of Molinet's poetry is not truly individualized. In narrative texts, the *acteur* is normally a passive witness, whose references to the writing process occupy a limited, marginal position, and whose relationship to the historical author is equivocal. The narrator of the *Naufrage*, and his visual representation in manuscript *E*, constitute a typical instance. Later pieces, such as the *Gaiges*, elicit more vertical readings, as Brown has noted: Molinet signs his work more explicitly, and adopts a more prominent narratorial voice, which seems to coincide with his own.[11] Symptomatic of this verticality are the two editions of the *Naissance*, which respond to Molinet's closing signature by acknowledging his authorship, the first pictorially, the second explicitly on the title-page. Until Molinet's latter years, however, relatively few elements in his poetry invite readers to relate the *je* to the historical *hors-texte*. These tendencies are manifested in the inconsistency with which poems are attributed in both manuscripts and print. The surviving manuscript anthologies devoted to Molinet's poetry were, perhaps significantly, produced posthumously, after printing had begun to promote the importance of the author function in the expectations of readers.

As printing develops, later *rhétoriqueurs* not only involve themselves more closely in publication, but also employ techniques which promote a more vertical reading of the *acteur*. In the work of Pierre Gringore, for example, Brown has identified an increasing coincidence of the first-person voice and the historical author.[12] This trend is also exemplified in editions of Lemaire's and Bouchet's poetry. Features which suggest an awareness of printing's implications, such as attributions on title-pages, and metadiscursive passages, emphasize the poetic persona's relationship to the world outside the text. Lemaire and Bouchet comment on their writing in liminary material, and refer to other works they have composed. Their texts increasingly include details which

[11] Ibid. 214–27. [12] Ibid. 228–45.

anchor the writing subject in a historical context: Lemaire reproduces at length the titles of the luminaries to whom he dedicates his works, while Bouchet often alludes to his job as *procureur*, and even gives the *acteur* his own name, as in the *Labirynth*. The rise of the author function, then, is accompanied by a developing verticality in the presentation of the first-person authorial voice, often reflected in the presentational characteristics of editions. The influence of printing upon this tendency is clear.

The evolution towards verticality, and the role played by printing in this respect, may be further elucidated through an examination of the work of Clément Marot, an approximate contemporary of the later *rhétoriqueurs* (his poetic activity spans the period 1512–44). As Brown has noted, Marot's characteristically 'personal' writing, and concerns with authority over his work, are partly related to print.[13] Marot participated in the publication of much of his work, notably through a co-operative relationship with the Lyons publisher Étienne Dolet. In paratextual features such as title-pages and liminary pieces, he adopted a defensive, proprietorial stance towards his poetry, underlining his own editorial activity and criticizing the publishers of unauthorized editions, which were potentially dangerous to him in view of the heterodox religious opinions sometimes expressed in his verse. Such attitudes are clearly exhibited in a letter to Dolet, published among the liminary material in Dolet's 1538 edition of Marot's *Œuvres*. Here, the poet attacks unscrupulous printers who had incorporated seditious works, erroneously attributed to him, in their editions of his poetry. The present edition, by contrast, is authorized, authentic, and contains a wealth of new material certain to tempt prospective buyers:

> Si ay jetté hors de mon Livre non seulement les maulvaises, mais les bonnes choses, qui ne sont à moy, ne de moy: me contentant de celles, que nostre Muse nous produict. Toutesfois au lieu des choses rejectées (affin que les Lecteurs ne se plaignent) je y ay mis douze fois aultant d'aultres Oeuvres miennes, par cy devant non imprimées: mesmement deux Livres de Epigrammes.[14]

What is more, Marot's influence on publication produces volumes whose material form reflects and reinforces a vertical presentation of the authorial persona. Marot's construction of

[13] On the following details, see Brown, *Poets, Patron, and Printers*, 247–54.

[14] Marot, *Œuvres*, i. 10. Further references will be incorporated into the text.

anthologies of his writing, especially of his early verse as the *Adolescence Clementine* (first published in 1532), encourages the tracing of a biographical narrative through his work. The liminary epistle which introduces the *Adolescence* explains that his 'Oeuvres de jeunesse' are followed by 'Ouvrages de meilleure trempe', but that 'l'Adolescence ira devant' (i. 17–18): readers are invited to follow a poetic development closely linked to Marot's biography.[15] Indeed, as Marot continues to write, later editions of the *Adolescence* are brought up to date: in 1533 or 1534, Marot's more recent work is added to the *Adolescence* under the title *Suite*, foregrounding the tendency for poems in these volumes to follow a chronological order. The autobiographical element in Marot's verse is especially evident in his *épîtres*, which form a significant part of his output. These appear transparently to refer to events in the historical author's life, as their titles often make explicit: *Epistre à monseigneur de Lorraine nouvellement venu à Paris, par laquelle Marot luy presente le premier livre translaté de la Metamorphose de Ovide*; *Marot prisonnier escript au roy, pour sa delivrance* (i. 293, 316).

Such features are reflected in the dominant formal and stylistic strategies in Marot's verse, a brief analysis of which enables the scope of Brown's discussion to be extended. Marot's early work often displays the densely worked language of *rhétoriqueur* poetry, most notably in the allegorical scenario of *Le Temple de Cupido*, and the *rimes équivoques* of the *Petite Epistre au roy* (i. 27–42, 87). However, his verse comes to be characterized less by allegory, obtrusive rhetoric, and explicit structure, and more by a carefully cultivated artlessness.[16] Poetic techniques are employed not to reveal Marot's craftsmanship, but to conceal it, creating an illusion of naturalness which focuses readers' attention upon the writing subject rather than form, and suggests that the poetry faithfully represents Marot's personality. Versification becomes simpler, tending towards decasyllabic couplets. Rigid formal prescriptions, which inherently foreground their conventionality, are often called into question by being combined, as such titles as *La Dixiesme Elegie en forme de ballade* (i. 249) indicate; more flexible and less obviously artificial genres, such as the *épître* and *épigramme*, are preferred. The

[15] On this and other issues addressed in this paragraph, see G. Defaux, 'Clément Marot: Poésie, autobiographie et roman'.

[16] See A. Williams, *Clément Marot: Figure, Text and Intertext* (Lampeter, 1990), 145–7; Marot, *Œuvres*, i, pp. cxxxvii–cxlii; Defaux, 'Clément Marot'; LeBlanc, *Va lettre va*, ch. 4.

je often interrupts himself while speaking, conveying an impression of spontaneity, as in *Epistre au roy, du temps de son exil à Ferrare* (ii. 80–6). Here Marot, proclaiming to Francis I that he is no Lutheran, digresses into a prayer, before remembering whom he is addressing:

> Que dys je? Où suis je? O noble Roy Françoys,
> Pardonne moy, car ailleurs je pensoys. (ii. 83, vv. 121–2)

On close examination, the digression is far from artless; it belongs to a series of passages which flatter Francis by tacitly comparing him to various irreproachable rulers, such as Augustus and Louis XII (ii. 82, vv. 55–6; 85–6, vv. 201–4).[17] Nevertheless, it suggests that Marot is putting his thoughts directly into the poem; the gap between historical author and assumed voice is minimized.

Referential readings are also solicited by the use of proper names and deixis to create an *effet de réel*, thereby implying that textual elements refer to an extratextual reality.[18] In this respect, Marot's verse resembles that of Villon, which Marot famously edited.[19] Marot not only names himself frequently in his poems, but is also apt to mention real people and places. Poems are written 'En la prison claire, & nette de Chartres' (ii. 19, v. 9), or 'Vivant en paix dedans Venise' (ii. 111, v. 220); they refer to 'Vignals' (i. 326, v. 5), or 'le Roy Françoys' (ii. 31, v. 405). Whether his addressees are named or not, Marot often foregrounds his relationships to them through deictic elements such as 'I', 'you', 'here', or 'this', which locate a speaking subject in his utterance, and set that utterance in a spatio-temporal context. Hence in *A la royne de Navarre* (ii. 118–23), he alludes to his past relationship with Marguerite, thereby focusing readers' attention on a historical situation:

> Tu sçais comment, par parolles mutines
> Des envieux aux langues serpentines,
> Je fus contrainct (bien t'en peult souvenir)
> Par devers toy en franchise venir. (ii. 119, vv. 27–30)

[17] See R. Griffin, *Clément Marot and the Inflections of Poetic Voice* (Berkeley, 1974), 53–4.

[18] See R. Barthes, 'L'Effet de réel', 88.

[19] See Lazard, 'Clément Marot'; Dop-Miller, 'Clément Marot'. These observations, indeed, are indebted to three ground-breaking studies of Villon: N. F. Regalado, 'La Fonction poétique des noms propres dans le *Testament* de François Villon'; '*Effet de réel, effet du réel*: Representation and Reference in Villon's *Testament*'; '"En l'an de mon trentiesme aage": Date, Deixis, and Moral Vision in Villon's *Testament*'.

Clearly, the speaking subject's presence in Marot is illusory. The *je* often adopts stances heavily indebted to literary tradition,[20] while intertextual echoes reveal that the persona is partly constructed through literary texts: in *A la royne de Navarre*, for instance, the evocations of misery in exile owe much to Ovid's *Epistulae ex Ponto*.[21] What is significant, however, is the way in which Marot's readers are invited to respond to the poems, to read them referentially. Marot develops techniques of self-presentation already practised by Bouchet, and to a lesser extent Lemaire. An awareness of these techniques not only enables us to nuance our understanding of Marot's relationship to *rhétoriqueur* poetry, but also illuminates the way in which the first-person subject evolves from a primarily horizontal to a primarily vertical function, under the influence of printing.

From a critical and literary-historical standpoint, then, my examination of *mise en livre* has permitted a fuller appreciation of the very techniques which are most distinctive and interesting in *rhétoriqueur* poetry, the exploration of the possibilities offered by versification and figurative language; this is especially important because it is print which has helped render the *rhétoriqueurs*' aesthetic so foreign to twentieth-century readers. Additionally, more than studies of single authors or genres, this approach permits the evolution of *rhétoriqueur* poetics to be properly grasped. Various features are common to these poets: a moral vision of literature as an essentially public activity; an artisanal treatment of language, embracing the vernacular as a worthy medium; the use of allegory and late-medieval verse forms; the acknowledgement of shared literary values, as manifested in poets' praise for their precursors and contemporaries.[22] Yet within these parameters, a development takes place towards a more *lisible* text, and a more 'vertical' presentation of the poetic persona.

In addition, this study engages with more general, theoretical issues. It illustrates the importance, in any branch of literary studies, of examining texts' material presentation; such considerations are particularly relevant in the context of contemporary

[20] See Scollen-Jimack, 'Marot'.

[21] See Marot, *Œuvres*, ii. 898–902.

[22] Jodogne, 'Les "Rhétoriqueurs"', 160–8; S. Bagoly, '"De mainctz aucteurs une progression": Un siècle à la recherche du Parnasse français', 94–107.

information technology, which promises to revolutionize the ways in which texts are produced and conceived. Current principles of intellectual property derive essentially from a print-based economy of textual production; the *Regnars* lawsuits mark an early stage in a process which culminated in the concepts of copyright and plagiarism as they are currently enshrined in law. The Internet, however, permits texts and images to be reproduced and transmitted with a freedom which renders impossible any attempt at regulation within the existing legal framework. As the production of electronic texts becomes increasingly important in professional life, it will require notions of intellectual property to be thoroughly revised.[23]

New technological systems also have implications for art, as they permit electronically stored texts and images to be recycled, and combined with each other and with 'real-world' images. This raises important questions of meaning and intertextuality, which, as Peter Wollen has observed, must be addressed in order to establish an aesthetic appropriate to these media:

> The first requirement is the development of a heterogeneous theory of meaning, open rather than closed, involving different types of sign, and bringing semantics together with hermeneutics, reference with metaphor. The second is a specific (formal) theory of intertextual meaning, the way in which re-contextualization changes meaning, the double, hybrid coding involved in quoting, plagiarizing, grafting and so on, the back and forth of meaning between texts. Both these projects entail a reconsideration of the logical form of meaning.[24]

Although conducted at a practical rather than theoretical level, my examination of the diverse material contexts of *rhétoriqueur* poetry has engaged with these issues. Illustrated manuscripts and editions enable heterogeneous combinations of signs, and the interactions between them, to be analysed. The processes of recontextualization, and its effects upon meaning, have been observed in various contexts. Stock woodcuts, for example, when reused by printers, may offer surprising visual glosses upon the texts they accompany. More generally, the different aspects of a single text—its versification, its historical or moral content, the prominence it accords its author—may be foregrounded to

[23] See Landow, *Hypertext*, 196–201.

[24] P. Wollen, *Raiding the Icebox: Reflections on Twentieth-Century Culture* (London, 1993), 67.

varying degrees, according to the priorities observed in the manuscripts or editions in which the text is transmitted. While not specifically linguistic or philosophical, my analyses may thus contribute to the eventual development of semantic theories which account for the productions of current and future information technology.

This study also has implications for reception theory, by combining various reader-centred approaches to literature, and by showing the extent to which literature is rooted in social practice. The study of literature in its physical form requires conjunctions of features to be analysed: book production and ownership, the form and content of texts, the ways in which books foreground varying elements of the texts they transmit. Consequently, in examining such conjunctions and in reflecting upon my critical methodology, I have found it necessary to adopt, and where necessary partially to synthesize, the disparate approaches which Susan Suleiman has identified within reader-oriented criticism:[25] the study of the text as a communicative act; a focus on texts' readability in terms of cultural codes; examinations of the phenomenology of the reading process; accounts of individual, historical readers' reactions; socio-historical analyses of the reading public; and enquiry into the nature and possibilities of interpretation. For instance, the use of such factual clues to reception as illustration, readers' annotations, and details of ownership has reduced the difficulties inherent in reconstructing the horizon of expectations of past readers. Similarly, my examination of the *Regnars*' illustrations has involved the assessment both of foxes' cultural coding, which helps readers interpret the images, and of the processes by which readers construct meaning from the information successively provided in the woodcuts and chapters. Hence the relationship between literature and the book constitutes a privileged field for continuing reflection upon reception theory. Through linking diverse approaches, my study has enabled them to illuminate each other; historicist and theoretical methods, analyses of texts' literary and social contexts, have been fruitfully combined. Though such currents cannot entirely be reconciled, I have shown that they are much more complementary than is often supposed.

[25] Suleiman, 'Introduction'.

On a more specific level, my examination of book form's influence on reception enables two important modifications to be made to Jauss's model of the aesthetics of reception. First, Jauss's central notion of the reader's horizon of expectations must be extended to include expectations regarding the physical presentation of literary texts. Readers' literary and bibliographical expectations influence each other: specific literary forms are often presented in particular ways on the page, ways which audiences come to assimilate to their horizon of expectations. In early sixteenth-century France, vernacular didactic poetry was generally printed in octavo volumes, in bastard type, with little attention to layout; a larger format, or a different typeface, might foreground particular aspects of a text more than would a conventional presentation. Secondly, the role of book producers in mediating between author and public must be recognized. Jauss's conception of literary history, as a narrative of interaction between writer and reader,[26] is overly simplistic in this respect, and fails to account for the ways in which the roles of 'writer' and 'reader' may be adopted and combined at all stages of a text's existence.[27] In manuscript culture, for example, scribes, editors, compilers, *remanieurs*, and continuators engage in both readerly and writerly activity, in varying proportions. Textual producers must be inserted into Jauss's 'triangle of author, work, and public',[28] and new sets of expectations and interpretations considered. Scribes or printers initially receive a text in the light of their horizon of expectations. Both this and their normal working practices influence the form in which they transmit the text to a wider audience. In turn, the text's reception by the audience partly depends on its physical presentation, which may appeal to particular expectations. Hence in the Janot/Lotrian edition of Lemaire's *Temple d'Honneur et de Vertus*, a vogue for pastoral poetry, and the availability of appropriate woodblocks, influence an illustrative programme which foregrounds the piece's pastoral elements. Future studies of the historicity of literature must accommodate questions of book production in this fashion.

[26] H. R. Jauss, *Toward an Aesthetic of Reception*, tr. T. Bahti (Brighton, 1982), 15.
[27] See Chartier, 'Du livre', 104–5.
[28] Jauss, *Aesthetic of Reception*, 19.

Ultimately, my combination of literary and bibliographical analysis points towards the possibility of a new paradigm of literary study. The work of the anthropologist Pierre Bourdieu provides a fruitful means of conceptualizing this paradigm. Bourdieu has identified three modes of knowledge of the social world: an unreflective, 'phenomenological' knowledge of lived experience; an 'objectivist' knowledge of the social structures which make possible that experience; and a *connaissance praxéologique*, whose object is not only social structures but also the relationships between them and the social practices which reproduce them, a knowledge which treats practices as processes rather than products, 'en se situant dans le mouvement même de leur effectuation'.[29] He illustrates the inadequacy of objectivist knowledge, and the necessity for *connaissance praxéologique*, through a critique of Saussurean linguistics, whose privileging of *langue* over *parole* renders it impossible to integrate aspects of linguistic execution into a theoretical framework.[30] This epistemological model may be applied to the study of literature, for the *langue/parole* distinction is analogous to that between *texts*, linguistic abstractions which transcend the material, and *books*, the particular concrete actualizations of texts. As I have shown, literature cannot adequately be studied solely in terms of texts, for it is through books that texts are transmitted and perceived. As in anthropology and linguistics, objectivist knowledge is valuable, but in itself insufficient. It is important to examine not only textual structures, but also the ways in which these structures are manifested in the practice of textual production and transmission. A *connaissance praxéologique* of literature would enrich the knowledge of texts by addressing their presentation in books.[31]

My study's various implications for the history and theory of literature are perhaps ironic: the *rhétoriqueurs*, generally regarded as minor artists epitomizing inauthentic artifice, have proved valuable in helping to elucidate the most fundamental processes of literature, reading and writing. Such ironies indicate a final parallel

[29] P. Bourdieu, *Esquisse d'une théorie de la pratique, précédée de trois études d'ethnologie kabyle* (Geneva, 1972), 163.

[30] Ibid. 164–70.

[31] For a similar discussion of the literary implications of *connaissance praxéologique*, see Armstrong, 'Practice', 270.

between literary studies and Kadare's Palace of Dreams. In both institutions, the surface prestige of primary texts bears no necessary relation to the utility of the interpretations they elicit: 'It was even said that many Master-Dreams had come from the most undistinguished of files.'[32]

[32] Kadare, *Palace of Dreams*, 76.

BIBLIOGRAPHY

MANUSCRIPTS

Arnhem, Bibliotheek, MS 79.
Arras, Bibliothèque Municipale, ms. 619.
Brussels, Bibliothèque Royale Albert Ier, ms. 21521–21531.
Brussels, Bibliothèque Royale Albert Ier, ms. 21551–21569.
Brussels, Bibliothèque Royale Albert Ier, ms. II, 2545.
Brussels, Bibliothèque Royale Albert Ier, ms. IV, 144.
Brussels, Bibliothèque Royale Albert Ier, ms. IV, 541.
Brussels, Bibliothèque Royale Albert Ier, ms. IV, 630/14.
Cambridge, Gonville and Caius College, MS 187:220.
Carpentras, Bibliothèque Inguimbertine, ms. 412.
Courtrai, Stedelijke Openbare Bibliotheek, ms. 366.
Glasgow, Hunterian Museum, MS 12 (S. 2. 2).
Mons, Archives de l'État, manuscrits divers 114.
Paris, Bibliothèque de l'Arsenal, ms. 3521.
Paris, Bibliothèque Nationale de France, Collection Dupuy, ms. 503.
Paris, Bibliothèque Nationale de France, ms. fr. 1642.
Paris, Bibliothèque Nationale de France, ms. fr. 1683.
Paris, Bibliothèque Nationale de France, ms. fr. 1721.
Paris, Bibliothèque Nationale de France, ms. fr. 2159.
Paris, Bibliothèque Nationale de France, ms. fr. 2200.
Paris, Bibliothèque Nationale de France, ms. fr. 2375.
Paris, Bibliothèque Nationale de France, ms. fr. 14980.
Paris, Bibliothèque Nationale de France, ms. fr. 19165.
Paris, Bibliothèque Nationale de France, ms. fr. 22326.
Paris, Bibliothèque Nationale de France, ms. fr. 23988.
Paris, Bibliothèque Nationale de France, ms. fr. 24038.
Paris, Bibliothèque Nationale de France, ms. fr. 24052.
Paris, Bibliothèque Nationale de France, ms. fr. 25295.
Paris, Bibliothèque Nationale de France, ms. lat. 1428.
Paris, Bibliothèque Nationale de France, ms. Rothschild 471.
Paris, Bibliothèque Nationale de France, nouvelles acquisitions françaises 1541.
Paris, Bibliothèque Nationale de France, nouvelles acquisitions françaises 4061.
Paris, Bibliothèque Nationale de France, nouvelles acquisitions françaises 11555.
Paris, Bibliothèque Nationale de France, nouvelles acquisitions françaises 21532.
Poitiers, Médiathèque François-Mitterrand, ms. 440.
Saint-Omer, Bibliothèque Municipale, ms. 657.

St Petersburg, State M. E. Saltўkov-Shchedrin Public Library, MS Fr. F. v. XIV, 8.
Tournai, Bibliothèque Communale, ms. 105.
Vienna, Österreichische Nationalbibliothek, Cod. 2579.
Vienna, Österreichische Nationalbibliothek, Cod. 3441.

PRINTED BOOKS

For fifteenth- and sixteenth-century editions, single representative copies are cited. Texts by the authors studied are listed in the order of their first editions, the earliest first.

Molinet's works

Le Temple de Mars, n.d.o.p. [*c*.1476] (Chantilly, Musée Condé, IV, G. 15).
——Paris, Le Petit Laurens, n.d. [*c*.1491] (New York, Pierpont Morgan Library, PML 75124).
——Paris, Jean Trepperel, n.d. [*c*.1497–8] (BNF, Rés. Ye 1127).
——n.d.o.p. [Paris, Michel Le Noir, 1501–5] (BNF, Rés. Ye 220).
——Lyons, no pub. [Jean de Vingle], 18 December 1502 (BNF, Rothschild 2580).
——n.d.o.p. [Paris, Jean Trepperel, 1506–9] (BNF, Rés. Ye 273).
——Paris, no pub., n.d. [Alain Lotrian, *c*.1520] (BNF, Rés. Y^2 2579).
——Paris, no pub., n.d. [Alain Lotrian, 1526–30] (BNF, Rés. Ye 1282).
——Paris, no pub., n.d. [Alain Lotrian, *c*.1531] (Paris, Bibl. de l'École Nationale Supérieure des Beaux-Arts, Rés. Masson 469).
——in *Traictez singuliers*, Paris, Antoine Couteau for Galliot Du Pré, February 1525/6 (BNF, Rothschild 487).
——in *Traictez singuliers*, Paris, for Galliot Du Pré, 'February 1525' (Arsenal, Rés. 8° BL 8749).
La Resource du petit peuple, Valenciennes, no pub., n.d. [Jean de Liège, 1493–1500] (BR, LP 4281 A (facsimile)).
La Robe de l'archiduc, Valenciennes, Jean de Liège, n.d. [*c*.1500] (Musée Condé, IV, E. 89(2)).
Roman de la Rose moralisé, Paris, Anthoine Vérard, n.d. (BL, C. 22. c. 2).
La Tres desiree et prouffitable naissance de Charles d'Austrice, Valenciennes, Jean de Liège, n.d. [1500–3] (BNF, Rés. Ye 1077).
——n.d.o.p. [Lyons, Guillaume Balsarin, *c*.1503] (BNF, Rés. Ye 221).
La Complainte de Constantinoble, n.d.o.p. (BL, 11475 d. 31).
Le Testament de la Guerre, Lignan, no pub., n.d. (BNF, Rés. Ye 1376).
——n.d.o.p. [Paris, *c*.1520] (Musée Condé, IV, D. 92).
——Paris, widow Nicolas Buffet, 1559 (Aix-en-Provence, Bibl. Méjanes, Rés. S 44 (4)).

Les Faictz et dictz, Paris, Jean Longis and widow Jean Saint-Denis, 9 December 1531 (BNF, Rés. Ye 41).
——Paris, Jean Longis and others, 1537 (Oxford, Bodleian Library, Douce M 551).
——Paris, Alain Lotrian and others, 1540 (BNF, Rés. Ye 1340).
L'Art de rhétorique, in *Recueil d'arts de seconde rhétorique*, ed. E. Langlois (Paris, 1902), 214–52.
Chroniques, ed. G. Doutrepont and O. Jodogne (Brussels, 1935–7).
Les Faictz et dictz de Jean Molinet, ed. N. Dupire (Paris, 1936–9).

Lemaire's works

Le Temple d'Honneur et de Vertus, Paris, Anthoine Vérard, n.d. [1504] (Paris, Bibl. Mazarine, Rés. 21716).
——Paris, Michel Le Noir, 6 April 1504 (BNF, Rés. Ye 846).
——n.d.o.p. [Paris, Philippe Le Noir, *c.*1520] (BNF, Rés. Ye 859).
——Paris, Alain Lotrian and Denis Janot, n.d. [1530–4] (BL, C. 39. b. 14).
——in *Les Illustrations de Gaule et singularités de Troie*, Paris, Jean Réal for Jean Bonfons, 1548–9 (BNF, Rés. La^2 12).
——ed. H. Hornik (Geneva, 1957).
Les Chansons de Namur, Antwerp, Henri Heckert, October 1507 (BNF, Rés. Ye 942 (facsimile)).
La Pompe funeralle des obseques du feu roy dom Phelippes, n.d.o.p. [Antwerp, Willem Vorsterman, from 15 February 1508] (BNF, Rothschild 2432).
La Concorde du genre humain, Brussels, Thomas van der Noot, January 1508/9 (BNF, Rés. Ye 1043).
——ed. P. Jodogne (Brussels, 1964).
Épître du roi Louis XII à Hector de Troie, Paris, for Geoffroy de Marnef, August 1513 (Lyons, BM, Rés. 104 686).
——Paris, for Geoffroy de Marnef, August 1513 (Lyons, BM, Rés. 317 051).
——Paris, [Nicolas Higman] for Geoffroy de Marnef, July 1516 (BL, G. 10249 (6)).
——Paris, [Nicolas Higman] for Geoffroy de Marnef, 'July 1516' (Oxford, Codrington Library, i i 5. 7).
——Paris, for Jean and Enguilbert de Marnef and Pierre Viart, August 1519 (BL, 634. k. 6).
——Paris, for Jean and Enguilbert de Marnef and Pierre Viart, August 1521 (BNF, Rés. La^2 5 (5)).
——Paris, for François Regnault and Jean Petit, 1523 (BNF, Rés. La^2 6 (4)).
——in *Les Illustrations de Gaule et singularités de Troie*, Lyons, Jacques Mareschal, 1 June 1524 (Cambridge, Clare College, D'. 1. 30).
——in *Les Illustrations de Gaule et singularités de Troie*, Paris, Philippe Le Noir, July 1524 (BNF, Rés. La^2 8).

Épître du roi Louis XII à Hector de Troie, in *Les Illustrations de Gaule et singularités de Troie*, Lyons, Antoine du Ry, 1528 (BL, C. 38. g. 12).

——in *Les Illustrations de Gaule et singularités de Troie*, Paris, François Regnault, 1528 (BNF, Rés. La^2 8 A).

——in *Les Illustrations de Gaule et singularités de Troie*, Paris, Nicolas Higman for Ambroise Girault, 1533 (Cambridge, Trinity College, Grylls 6. 236).

——in *Les Illustrations de Gaule et singularités de Troie*, Paris, François Regnault [in association with Philippe Le Noir], n.d. [1533–4] (BNF, Rés. La^2 7).

——in *Les Illustrations de Gaule et singularités de Troie*, Lyons, Jean de Tournes, 1549 (Lyons, BM, Rés. 108 167).

Œuvres, ed. J. Stecher (Louvain, 1882–91).

La Plainte du Désiré, ed. D. Yabsley (Geneva, 1932).

La Concorde des deux langages, ed. J. Frappier (Paris, 1947).

Les Épîtres de l'Amant vert, ed. J. Frappier (Geneva, 1948).

Traicté de la différence des schismes et des conciles de l'Église, avec l''Histoire du Prince Sophy' et autres œuvres, ed. J. Britnell (Geneva, 1997).

Bouchet's works

Les Regnars traversant, Paris, Anthoine Vérard, n.d. [1503–4] (Bodleian, Douce B 787).

——Paris, Anthoine Vérard, n.d. (BNF, Vélins 1103).

——Paris, Michel Le Noir, 21 May 1504 (Bodleian, 4° Buchanan d. 36).

——Paris, Philippe Le Noir, 1522 (BSG, 4° Y. 426^{25} inv. 589 rés.).

——Paris, Philippe Le Noir for Denis Janot, 25 January 1530 (Bodleian, Douce B subt. 248).

L'Histoire et cronicque de Clotaire, Poitiers, Enguilbert de Marnef, n.d. [1518] (BNF, Rés. 4° L^4 b. 4).

Le Labirynth de Fortune, Poitiers, Enguilbert de Marnef and Jacques Bouchet, n.d. [*c.*1522] (Musée Condé, IV, E. 82).

——Poitiers, Enguilbert de Marnef and Jacques Bouchet, 26 March 1524 (Arsenal, 8° BL 8744).

——Paris, Alain Lotrian, n.d. (BNF, Rés. Ye 353).

——Paris, Philippe Le Noir, n.d. (Bodleian, Douce BB 507).

——Paris, Philippe Le Noir, n.d. (BNF, Rés. Ye 354).

Les Cantiques et oraisons contemplatives de l'ame penitente, gathering bound with *Heures à l'usaige de Poictiers*, Poitiers, Jehan Coussot for Jacques Bouchet and Nicolas Pelletier, 20 December 1525 (BNF, Vélins 1655).

Les Anciennes et modernes genealogies des roys de France, Poitiers, Jacques Bouchet, 26 January 1527/8 (MFM, CR. 152).

——Poitiers, Jacques Bouchet, 27 November 1529 (MFM, DP. 1781).

——Paris, no pub., 1530 (Tchemerzine, *Bibliographie*, ii. 51–2).

——Poitiers, Jacques Bouchet, 27 November 1531 (BL, 596. b. 11).

——Poitiers, Jacques Bouchet, 12 June 1535 (MFM, DM. 843 (1)).
——Paris, Galliot Du Pré, 1536 (MFM, DM. 844).
——Poitiers, Jacques Bouchet, 16 July 1537 (MFM, DP. 116).
——Paris, various booksellers, 1537 (MFM, DM. 846).
——Paris, various booksellers, 1539 (Bodleian, 8° S. 231 Art).
——Paris, various booksellers, 1541 (BNF, Rés. L^{37}. 2. E).
Les Triumphes de la noble et amoureuse dame, Poitiers, Jacques Bouchet, 20 June 1530 (BSG, Rés. Y fol. 141 inv. 201).
——Rennes, Jean Georget, 1541 (BNF, Vélins 586).
Le Jugement poetic de l'honneur femenin, Poitiers, Jean and Enguilbert de Marnef, 1 April 1538 (MFM, DP. 113).
Forme et ordre de plaidoirie, Poitiers, Jean and Enguilbert de Marnef, 1542 (MFM, DM. 1402 (1)).
Les Genealogies, effigies et epitaphes des roys de France, Poitiers, Jacques Bouchet and Jean and Enguilbert de Marnef, 1545 (Bodleian, Buchanan c. 19).
Triomphes du tres Chrestien Roy de France, Poitiers, Jean and Enguilbert de Marnef, 17 August 1549 (BNF, Rés. Ye 191).
Epistres morales et familieres du Traverseur, ed. J. Beard (Wakefield, 1969).
La Déploration de l'Église militante, ed. J. Britnell (Geneva, 1991).
Le Temple de Bonne Renommee, ed. G. Bellati (Milan, 1992).

Other literary texts

ANDRELINI, P. F., *Amores sive Livia*, ed. G. Tournoy-Thoen (Brussels, 1982).
BAEBIUS ITALICUS, *Ilias Latina*, ed. M. Scaffai (Bologna, 1982).
BRANT, S., *Varia carmina*, Basle, Johann Bergmann von Olpe, 1498 (Bodleian, Auct. 2 Q 6. 33).
——*Flugblätter*, ed. P. Heitz (Strasbourg, 1915).
CHARTIER, A., *Les Faictz et dictz*, Paris, Galliot Du Pré, 1526 (Bodleian, MS E. D. Clarke 33).
——*Poetical Works*, ed. J. C. Laidlaw (Cambridge, 1974).
CHASTELAIN, G., *Œuvres*, ed. K. de Lettenhove (Brussels, 1863–6).
CRETIN, G., *Le Debat de deux dames*, Paris, Antoine Couteau for Jean Longis, 1 April 1526 (BNF, Rés. Ye 1337).
——*Œuvres poétiques*, ed. K. Chesney (Paris, 1932).
Les Croniques des roys, ductz et contes de Bourgoigne, n.d.o.p. [Lyons, Guillaume Balsarin, *c.*1503] (BR, Inc. A 409 LP).
DARES PHRIGIUS, *De excidio Troie*, Paris, Nicolas Des Prez for Pierre Gaudoul, 1520 (BNF, Rés. J. 1719).
FABRI, P., *Le Grand et vrai art de pleine rhétorique*, ed. A. Héron (Rouen, 1890).
GOBIN, R., *Les Loups ravissans*, Paris, Anthoine Vérard, n.d. (BNF, Rés. Ye 851).
GRINGORE, P., *Les Folles Entreprises*, Paris, Pierre Le Dru, 23 December 1505 (Bodleian, Douce G 386).

GRINGORE, P., *Les Fantasies de Mere Sote*, ed. R. L. Frautschi (Chapel Hill, NC, 1962).

HORACE, *The Odes and Epodes*, ed. and tr. C. E. Bennett (Cambridge, Mass., 1927).

KADARE, I., *The Palace of Dreams*, tr. J. Vrioni, tr. B. Bray (London, 1993).

LA VIGNE, A. DE, *La Ressource de la Chrestienté*, ed. C. J. Brown (Montreal, 1989).

LORRIS, G. DE, and MEUN, J. DE, *Le Roman de la Rose*, ed. F. Lecoy (Paris, 1965–70).

Louenges a Nostre Seigneur, Paris, Anthoine Vérard, n.d. [*c.*1502] (BNF, Rés. Ye 831).

MACROBIUS, *Commentarii in somnium Scipioni libri duo*, ed. L. Scarpa (Padua, 1981).

MAROT, C., *Œuvres poétiques complètes*, ed. G. Defaux (Paris, 1990–3).

MAROT, J., *Recueil*, ed. A. Saunders (Menston, 1971).

MICHAULT, P., *Le Procès d'Honneur Féminin*, ed. B. Folkart, *MF* 2 (1978).

Les Motz dorez de Cathon, Paris, no pub., n.d. [Guillaume de Bossozel and Jean Longis, *c.*1530] (BNF, Rés. p. Yc. 906).

Les Neuf Preux, Paris, Michel Le Noir, 3 December 1507 (BL, C. 107. g. 6).

Ovid, *'The Art of Love', and Other Poems*, ed. and tr. J. H. Mozley, rev. G. P. Goold (2nd edn. Cambridge, Mass., 1979).

Le Petit Artus de Bretaigne, Lyons, Jean de la Fontaine, 4 July 1493 (ÖNB, Ink. 11. F. 35).

Recueil d'arts de seconde rhétorique, ed. E. Langlois (Paris, 1902).

Recueil de pièces historiques imprimées sous le règne de Louis XI, ed. É. Picot and H. Stein (Paris, 1923).

VIRGIL, *Œuvres*, Paris, Nicolas Couteau for Galliot Du Pré, 1529 (Arsenal, 4° BL 1782).

—— *P. Vergili Maronis opera*, ed. R. A. B. Mynors (Oxford, 1969).

Secondary literature

ABÉLARD, J., *'Les Illustrations de Gaule et singularitez de Troye' de Jean Lemaire de Belges: Étude des éditions—genèse de l'œuvre* (Geneva, 1976).

—— 'Les *Illustrations de Gaule* de Jean Lemaire de Belges: Quelle Gaule? Quelle France? Quelle nation?', *NRSS* 13 (1995), 7–27.

Académie Royale des Sciences, des Lettres et des Beaux-Arts de Belgique, *Biographie nationale* (Brussels, 1866–1985).

ADAMS, A., 'Le Manuscrit huntérien de l'*Epistre du roy* de Jean Lemaire de Belges et le manuscrit B.N. fonds français 1690', *BHR* 45 (1983), 511–18.

ANGELI, G., 'Il "rude engin" di Molinet', *Paragone*, 328 (June 1977), 21–47.

APEL, W., *The Notation of Polyphonic Music 900–1600* (5th edn. Cambridge, Mass., 1953).

ARMSTRONG, A., '*Grand Rhétoriqueur* Poetics: From Manuscript to Print', D.Phil. thesis (Oxford, 1995).

—— 'More Manuscript Copies of Jean Bouchet's Verse: Mss. B.N. fr. 2206 and 2231', *BHR* 57 (1995), 89–99.

—— 'The Practice of Textual Transmission: Jean Molinet's *Ressource du petit peuple*', *Forum for Modern Language Studies*, 33 (1997), 270–82.

——'Two More Rebus-Poems by Jean Molinet?' *Scriptorium*, 51 (1997), 76–80.

ARMSTRONG, E., *Before Copyright: The French Book-Privilege System 1498–1526* (Cambridge, 1990).

ATTWOOD, C., *Dynamic Dichotomy: The Poetic 'I' in Fourteenth- and Fifteenth-Century French Poetry* (Amsterdam, 1998).

AVRIL, F., and REYNAUD, N., *Les Manuscrits à peintures en France 1440–1520* (Paris, 1993).

BAGOLY, S., '"De mainctz aucteurs une progression": Un siècle à la recherche du Parnasse français', *MF* 17 (1985), 83–123.

BARTHES, R., 'Rhétorique de l'image', *Communications*, 4 (1964), 40–51.

——'L'Effet de réel', *Communications*, 11 (1968), 84–9.

——*S/Z* (Paris, 1970).

BATTERSBY, C., *Gender and Genius: Towards a Feminist Aesthetics* (London, 1989).

BAUDRIER, H., *Bibliographie lyonnaise* (Lyons, 1895–1921).

BAYOT, A., 'Notice du ms. original des *Mémoires* de Jean de Haynin', *Revue des bibliothèques et archives de Belgique*, 6 (1908), 109–44.

BEARD, J., 'Letters from the Elysian Fields: A Group of Poems for Louis XII', *BHR* 31 (1969), 27–38.

BELLATI, G., 'Étude sur les *Regnars traversant* de Jean Bouchet', in *Les Grands Rhétoriqueurs*, i (Milan, 1985), 125–52.

——'Bouchet e l'abolizione della *coupe féminine*', *Studi francesi*, 94 (1988), 73–81.

BENVENISTE, É., 'Structure des relations de personne dans le verbe', *Bulletin de la Société de Linguistique de Paris*, 43.1 (1946), 1–12.

——'La Nature des pronoms', in M. Halle, H. G. Lunt, H. McLean, and C. H. van Schooneveld (eds.), *For Roman Jakobson: Essays on the Occasion of his Sixtieth Birthday* (The Hague, 1956), 34–7.

BERGWEILER, U., *Die Allegorie im Werk von Jean Lemaire de Belges* (Geneva, 1976).

Bibliothèque Royale Albert Ier (Brussels), *Quinze années d'acquisition: De la pose de la première pierre à l'inauguration officielle de la Bibliothèque* (Brussels, 1969).

BLOCKMANS, W. P. (ed.), *Le Privilège général et les privilèges régionaux de Marie de Bourgogne pour les Pays-Bas, 1477* (Courtrai, 1985).

BLOOM, H., *A Map of Misreading* (Oxford, 1975).

BOURDIEU, P., *Esquisse d'une théorie de la pratique, précédée de trois études d'ethnologie kabyle* (Geneva, 1972).

BRITNELL, J., 'Jean Bouchet's Prayers in French for the Laity, *Les Cantiques et oraisons contemplatives de l'âme pénitente traversant les voies périlleuses*', *BHR* 38 (1976), 421–36.

——*Jean Bouchet* (Edinburgh, 1986).

——'The Antipapalism of Jean Lemaire de Belges' *Traité de la difference des schismes et des conciles*', *Sixteenth Century Journal*, 24 (1993), 783–800.

——'La Mort de Jean Lemaire de Belges, l'édition de 1517 du *Traité des schismes et des conciles*, et les impertinences d'un éditeur', *BHR* 56 (1994), 127–33.

BROWN, C. J., 'Jean Lemaire's *La Concorde des deux langages*: The Merging of Politics, Language, and Poetry', *Fifteenth-Century Studies*, 3 (1980), 29–39.

BROWN, C. J., 'The Rise of Literary Consciousness in Late Medieval France: Jean Lemaire de Belges and the *Rhétoriqueur* Tradition', *JMRS* 13 (1983), 51–74.

——'Du manuscrit à l'imprimé en France: Le Cas des Grands Rhétoriqueurs', in *Les Grands Rhétoriqueurs*, i (Milan, 1985), 108–23.

——*The Shaping of History and Poetry in Late Medieval France: Propaganda and Artistic Expression in the Works of the Rhétoriqueurs* (Birmingham, Ala., 1985).

——'L'Éveil d'une nouvelle conscience littéraire en France à la grande époque de transition technique: Jean Molinet et son moulin poétique', *MF* 22 (1989), 15–35.

——'Text, Image, and Authorial Self-Consciousness in Late Medieval Paris', in S. L. Hindman (ed.), *Printing the Written Word* (Ithaca, NY, 1991), 103–42.

——*Poets, Patrons, and Printers: Crisis of Authority in Late Medieval France* (Ithaca, NY, 1995).

BÜHLER, C. F., *The Fifteenth-Century Book: The Scribes, the Printers, the Decorators* (Philadelphia, 1960).

CAMILLE, M., 'Reading the Printed Image: Illuminations and Woodcuts of the *Pèlerinage de la vie humaine* in the Fifteenth Century', in S. L. Hindman (ed.), *Printing the Written Word* (Ithaca, NY, 1991), 259–90.

CAMPANGNE, H., *Mythologie et rhétorique aux XVe et XVIe siècles en France* (Paris, 1996).

CARTIER, A., *Bibliographie des éditions des de Tournes, imprimeurs lyonnais* (Paris, 1937–8).

CASTOR, G., *Pléiade Poetics: A Study in Sixteenth-Century Thought and Terminology* (Cambridge, 1964).

CERQUIGLINI, B., *Éloge de la variante: Histoire critique de la philologie* (Paris, 1989).

CERQUIGLINI, J., 'L'Écriture louche: La Voie oblique chez les Grands Rhétoriqueurs', in *Les Grands Rhétoriqueurs*, i (Milan, 1985), 21–31.

CHARON-PARENT, A., 'Aspects de la politique éditoriale de Galliot Du Pré', in P. Aquilon and H.-J. Martin (eds.), *Le Livre dans l'Europe de la Renaissance: Actes du XXVIIIe colloque international d'études humanistes de Tours* (Paris, 1988), 209–18.

CHARTIER, R., *L'Ordre des livres: Lecteurs, auteurs, bibliothèques en Europe entre XIVe et XVIIIe siècle* (Aix-en-Provence, 1992).

——'Du livre au lire', in R. Chartier (ed.), *Pratiques de la lecture* (2nd edn. Paris, 1993), 79–113.

Le Cinquième Centenaire de l'imprimerie dans les anciens Pays-Bas (exposition à la Bibliothèque Royale Albert Ier, du 11 septembre au 27 octobre, 1973): Catalogue (Brussels, 1973).

CLIER-COLOMBANI, F., *La Fée Mélusine au Moyen Âge: Images, mythes et symboles* (Paris, 1991).

COCKSHAW, P., LEMAIRE, C., and ROUZET, A., *Charles le Téméraire: Exposition organisée à l'occasion du cinquième centenaire de sa mort* (Brussels, 1977).

COLIN, G., 'Jean de Liège', in *Le Cinquième Centenaire de l'imprimerie dans les anciens Pays-Bas* (Brussels, 1973), 529–33.

——'L'Imprimeur du Flavius Josèphe', in *Le Cinquième Centenaire de l'imprimerie dans les anciens Pays-Bas* (Brussels, 1973), 181–4.

COLLARD, F., 'Une œuvre historiographique du règne de Charles VIII et sa réception: Le *Compendium de origine et gestis Francorum* de Robert Gaguin', *NRSS* 13 (1995), 71–86.

CONLEY, T., *The Graphic Unconscious in Early Modern French Writing* (Cambridge, 1992).

COOPER, H., *Pastoral: Mediaeval into Renaissance* (Ipswich, 1977).

COOPER, R., *Litterae in tempore belli: Études sur les relations littéraires italo-françaises pendant les guerres d'Italie* (Geneva, 1997).

CORNILLIAT, F., 'Aspects du songe chez les derniers rhétoriqueurs: Analyse du *Labirynth de Fortune* et du *Séjour d'Honneur*', *Revue d'humanisme et Renaissance*, 13 (1987), 17–37.

—— *'Or ne mens': Couleurs de l'éloge et du blâme chez les 'Grands Rhétoriqueurs'* (Paris, 1994).

COTGRAVE, R., *A Dictionarie of the French and English Tongues* (London, 1611).

COUTURIER, M., *Textual Communication: A Print-Based Theory of the Novel* (London, 1991).

COWLING, D., *Building the Text: Architecture as Metaphor in Late Medieval and Early Modern France* (Oxford, 1998).

CULLER, J., *Structuralist Poetics: Structuralism, Linguistics and the Study of Literature* (London, 1975).

CURTIUS, E. R., 'Der Archipoeta und der Stil mittellateinischer Dichtung', *Romanische Forschungen*, 54 (1940), 105–64.

—— *Europäische Literatur und lateinisches Mittelalter* (2nd edn. Berne, 1954).

DAVIES, H. W., *Catalogue of a Collection of Early French Books in the Library of C. Fairfax Murray* (London, 1910).

DEBAE, M., *La Librairie de Marguerite d'Autriche* (Brussels, 1987).

DEFAUX, G., 'Clément Marot: Poésie, autobiographie et roman', in R. C. La Charité (ed.), *Writing the Renaissance: Essays on Sixteenth-Century French Literature in Honor of Floyd Gray* (Lexington, Ky., 1992), 79–91.

DE LOOZE, L., *Pseudo-autobiography in the Fourteenth Century: Juan Ruiz, Guillaume de Machaut, Jean Froissart, and Geoffrey Chaucer* (Gainesville, Fla., 1997).

DEVAUX, J., 'Le Rôle politique de Marie de Bourgogne au lendemain de Nancy: Vérité ou légende?' *MA* 97 (1991), 389–405.

—— 'Les Soulèvements urbains de 1477 sous le regard des chroniqueurs du temps', in *Actes du LIe congrès de la Fédération des Cercles d'Archéologie et d'Histoire de Belgique*, ii (Liège, 1994), 391–411.

—— 'Les Guerres de Louis XI au *pays de Haynau*: Le Témoignage de Jean Molinet', *Annales du Cercle Archéologique de Mons*, 76 (1994), 95–113.

—— *Jean Molinet, indiciaire bourguignon* (Faris, 1996).

DOOB, P. R., *The Idea of the Labyrinth from Classical Antiquity through the Middle Ages* (Ithaca, NY, 1990).

DOP-MILLER, C., 'Clément Marot et l'édition humaniste des œuvres de François Villon', *Romania*, 112 (1991), 217–42.

DUBOIS, C.-G., 'L'Invention littéraire et les jeux de langage: Jeux de nombres, jeux de sons, jeux de sens', in P. Ariès and J.-C. Margolin (eds.), *Les Jeux à la Renaissance: Actes du XXIIIe colloque international d'études humanistes, Tours, juillet 1980* (Paris, 1982), 245–69.

DUPIRE, N., *Jean Molinet: La Vie—les œuvres* (Paris, 1932).

—— *Étude critique des manuscrits et éditions des poésies de Jean Molinet* (Paris, 1932).

EISENSTEIN, E., *The Printing Press as an Agent of Change: Communications and Cultural Transformations in Early-Modern Europe* (Cambridge, 1979).

ERLICH, V., *Russian Formalism: History—Doctrine* (The Hague, 1955).

FAIDER, P., and VAN SINT JAN, P., *Catalogue général des manuscrits des bibliothèques de Belgique*, vi: *Catalogue des manuscrits conservés à Tournai (Bibliothèques de la Ville et Séminaire)* (Gembloux, 1950).

FEBVRE, L., and MARTIN, H.-J., *L'Apparition du livre* (Paris, 1958).

FENOALTEA, D., 'Doing It with Mirrors: Architecture and Textual Construction in Jean Lemaire's *La Concorde des deux langages*', in B. C. Bowen and J. C. Nash (eds.), *Lapidary Inscriptions: Renaissance Essays for Donald A. Stone, Jr.* (Lexington, Ky., 1991), 21–32.

FLINN, J., *Le 'Roman de Renart' dans la littérature française et dans les littératures étrangères au Moyen Âge* (Toronto, 1963).

FOUCAULT, M., 'Qu'est-ce qu'un auteur?', *Bulletin de la Société Française de Philosophie*, 63 (1969), 73–104.

FRAPPIER, J., 'Nouvelles Réflexions sur *La Concorde des deux langages*', *BFLUS* 28 (1948–9), 1–12.

FREUD, S., *The Interpretation of Dreams* (London, 1954).

GARNIER, F., *Le Langage de l'image au Moyen Âge* (Paris, 1982–8).

GASKELL, P., *A New Introduction to Bibliography* (Oxford, 1972).

GENETTE, G., *Seuils* (Paris, 1987).

GIARD, R., and LEMAÎTRE, H., 'Les Origines de l'imprimerie à Valenciennes: Jehan de Liège', *Bulletin du bibliophile et du bibliothécaire* (1903), 349–62.

GODEFROY, F., *Dictionnaire de l'ancienne langue française* (Paris, 1881–1902).

GOMBRICH, E., *Art and Illusion: A Study in the Psychology of Pictorial Representation* (5th edn. Oxford, 1977).

GOODMAN, K. S., *Language and Literacy: The Selected Writings of Kenneth S. Goodman*, ed. F. V. Gollasch (London, 1982).

GORDON, A. L., '*La Ressource du petit peuple* (1481): Essai de pleine rhétorique', *Travaux de littérature*, 2 (1989), 55–67.

Les Grands Rhétoriqueurs: Actes du 5e colloque sur le moyen français, Milan, 6–8 mai 1985, i (Milan, 1985).

GRIFFIN, R., *Clément Marot and the Inflections of Poetic Voice* (Berkeley, 1974).

—— '*La Concorde des deux langages: Discordia concors*', in P. M. Smith and I. D. Macfarlane (eds.), *Literature and the Arts in the Reign of Francis I: Essays Presented to C. A. Mayer* (Lexington, Ky., 1985), 54–81.

Gros, G., '*Querant l'un œil envers les cieulx estendre* . . . : Étude sur la *Louenge* mariale de George Chastelain', *MA* 98 (1992), 429–45.

——*Le Poète marial et l'art graphique: Étude sur les jeux de lettres dans les poèmes pieux du Moyen Âge* (Caen, 1993).

——'Sur l'oraison mariale de Valenciennes, ou Jean Lemaire de Belges et ses devanciers', in J.-C. Aubailly, E. Baumgartner, F. Dubost, L. Dulac, and M. Faure (eds.), *'Et c'est la fin pour quoy sommes ensemble': Hommage à Jean Dufournet* (Paris, 1993), ii. 685–93.

Groupe µ, *Traité du signe visuel* (Paris, 1992).

Guy, H., *Histoire de la poésie française au XVIe siècle*, i: *L'École des rhétoriqueurs* (Paris, 1910).

Hasenohr, G., 'Traductions et littérature en langue vulgaire', in H.-J. Martin and J. Vezin (eds.), *Mise en page et mise en texte du livre manuscrit* (Paris, 1990), 229–352.

Hassig, D., *Medieval Bestiaries: Text, Image, Ideology* (Cambridge, 1995).

Hawkes, T., *Structuralism and Semiotics* (London, 1977).

Hindman, S. L. (ed.), *Printing the Written Word: The Social History of Books, circa 1450–1520* (Ithaca, NY, 1991).

Hirsch, R., *Printing, Selling and Reading 1450–1550* (Wiesbaden, 1967).

——'Title Pages in French Incunables, 1486–1500', *Gutenberg-Jahrbuch* (1978), 63–6.

Holub, R. C., *Reception Theory: A Critical Introduction* (London, 1984).

Hüe, D., 'Enjeux et emblèmes de la poésie à lectures multiples', *Romania*, 109 (1988), 386–95.

Hult, D., 'Reading It Right: The Ideology of Text Editing', in M. S. Brownlee, K. Brownlee, and S. G. Nichols (eds.), *The New Medievalism* (Baltimore, 1991), 113–30.

Hulubei, A., *L'Églogue en France au XVIe siècle: Époque des Valois (1515–1589)* (Paris, 1938).

Huot, S., *From Song to Book: The Poetics of Writing in Old French Lyric and Lyrical Narrative Poetry* (Ithaca, NY, 1987).

——*The 'Romance of the Rose' and its Medieval Readers: Interpretation, Reception, Manuscript Transmission* (Cambridge, 1993).

Hutcheon, L., *Narcissistic Narrative: The Metafictional Paradox* (London, 1984).

Imbs, P., 'Jean Lemaire de Belges: *La Concorde des deux langages*', *BFLUS* 26 (1947–8), 181–7.

Ivins, W. M., *Prints and Visual Communication* (London, 1953).

Jakobson, R., 'What Is Poetry?', tr. M. Heim, in L. Matejka and I. R. Titunik (eds.), *Semiotics of Art: Prague School Contributions* (Cambridge, Mass., 1976), 164–75.

Jauss, H. R., *Toward an Aesthetic of Reception*, tr. T. Bahti (Brighton, 1982).

Jodogne, P., 'Un recueil poétique de J. Lemaire en 1498', in F. Simone (ed.), *Miscellanea di studi e ricerche sul Quattrocento francese* (Turin, 1967), 179–210.

JODOGNE, P., 'Les "Rhétoriqueurs" et l'humanisme: Problème d'histoire littéraire', in A. H. T. Levi (ed.), *Humanism in France at the End of the Middle Ages and in the Early Renaissance* (Manchester, 1970), 150–75.

——*Jean Lemaire de Belges, écrivain franco-bourguignon* (Brussels, 1972).

——'Le Fragment manuscrit Bruxelles, B.R., IV 630/14: Une page des *Illustrations de Gaule* de Jean Lemaire de Belges', *Scriptorium*, 38 (1984), 101–4.

JOHNSON, L. W., *Poets as Players: Theme and Variation in Late Medieval French Poetry* (Stanford, Calif., 1990).

JUNG, M.-R., 'Poetria: Zur Dichtungstheorie des ausgehenden Mittelalters in Frankreich', *Vox Romanica*, 30–1 (1971), 44–64.

——'Les *Douze dames de rhétorique*', in P. Wunderli (ed.), *Du mot au texte: Actes du III*[ème] *colloque international sur le moyen français, Düsseldorf, 17–19 septembre 1980* (Tübingen, 1982), 229–40.

——'Simon Gréban, rhétoriqueur', in S. Cigada and A. Slerca (eds.), *Actes du VI*[e] *colloque international sur le moyen français, Milan, 4–6 mai 1988* (Milan, 1991), ii. 57–85.

KEM, J., 'Allegorical Images of the Cardinal Virtue Prudence in the Works of the *Rhétoriqueurs*', *RN* 28 (1987/8), 177–85.

——*Jean Lemaire de Belges's 'Les Illustrations de Gaule et singularitez de Troye': The Trojan Legend in the Late Middle Ages and Early Renaissance* (New York, 1994).

KENNEDY, E., 'The Scribe as Editor', in *Mélanges de langue et de littérature du Moyen Âge et de la Renaissance offerts à Jean Frappier* (Geneva, 1970), i. 523–31.

KENNEDY, W., 'Petrarchan Audiences and Print Technology', *JMRS* 14 (1984), 1–20.

KIRSCHBAUM, E. (ed.), *Lexikon der christlichen Ikonographie* (Rome, 1968–76).

KITTAY, E. F., *Metaphor: Its Cognitive Force and Linguistic Structure* (Oxford, 1987).

KNECHT, R. J., *Renaissance Warrior and Patron: The Reign of Francis I* (Cambridge, 1994).

LABANDE-MAILFERT, Y., *Charles VIII et son milieu (1470–1498): La Jeunesse au pouvoir* (Paris, 1975).

LABARRE, A., *Le Livre dans la vie amiénoise du seizième siècle: L'Enseignement des inventaires après décès, 1503–1576* (Paris, 1971).

LABORDE, Comte A. DE, *Les Principaux Manuscrits à peintures conservés dans l'ancienne Bibliothèque Impériale Publique de Saint-Pétersbourg* (Paris, 1938).

LA BORDERIE, A. DE, *Archives du bibliophile breton* (Rennes, 1880–1907).

LACHÈVRE, F., *Les Recueils collectifs de poésies libres et satiriques publiés depuis 1600 jusqu'à la mort de Théophile (1626)* (Paris, 1914).

LANDOW, G. P., *Hypertext: The Convergence of Contemporary Critical Theory and Technology* (Baltimore, 1992).

LAZARD, M., 'Clément Marot éditeur et lecteur de Villon', *CAIEF* 32 (1980), 7–20.

LEBLANC, Y., *Va lettre va: The French Verse Epistle (1400–1550)* (Birmingham, Ala., 1995).

LEMAIRE, J., 'Deux poèmes bourguignons sur la prise d'Arras en 1492: Un inédit de Molinet?', *Revue du Nord*, 60 (1978), 57–64.

—— 'Meschinot, Molinet, Villon: Témoignages inédits. Étude du Bruxellensis IV 541, suivie de l'édition de quelques ballades', *Archives et bibliothèques de Belgique*, special issue, 20 (1979).

—— *Introduction à la codicologie* (Louvain-la-Neuve, 1989).

LENNARD, J., *The Poetry Handbook: A Guide to Reading Poetry for Pleasure and Practical Criticism* (Oxford, 1996).

LOSSE, D. N., *Sampling the Book: Renaissance Prologues and the French 'Conteurs'* (Lewisburg, Pa., 1994).

LOUANT, A., *Le 'Livre de ballades' de Jehan et Charles Bocquet, bourgeois de Mons au XVI*[e] *siècle* (Brussels, 1954).

LUBORSKY, R. S., 'Connections and Disconnections between Images and Texts: The Case of Secular Tudor Book Illustration', *Word and Image*, 3 (1987), 74–85.

LYDGATE, B., 'Printing, Narrative, and the Genesis of the Rabelaisian Novel', *Romanic Review*, 71 (1980), 345–73.

—— 'Mortgaging One's Work to the World: Publication and the Structure of Montaigne's *Essais*', *Publications of the Modern Language Association of America*, 96 (1981), 210–23.

MACFARLANE, J., *Antoine Vérard* (London, 1900).

MCGANN, J. J., *A Critique of Modern Textual Criticism* (Chicago, 1983).

—— *The Textual Condition* (Princeton, 1991).

MCHALE, B., *Postmodernist Fiction* (London, 1987).

MARTIN, H.-J., and CHARTIER, R. (eds.), *Histoire de l'édition française*, i: *Le Livre conquérant* (Paris, 1982).

—— and VEZIN, J. (eds.), *Mise en page et mise en texte du livre manuscrit* (Paris, 1990).

MAYER, C. A., and BENTLEY-CRANCH, D., 'Florimond Robertet: Italianisme et Renaissance française', in *Mélanges à la mémoire de Franco Simone*, iv: *Tradition et originalité dans la création littéraire* (Geneva, 1983), 135–49.

MÉCHOULAN, É., 'Les Arts de rhétorique du XVe siècle: La Théorie, masque de la *theoria*?', in M.-L. Ollier (ed.), *Masques et déguisements dans la littérature médiévale* (Montreal, 1988), 213–21.

MISKIMIN, A., *The Renaissance Chaucer* (New Haven, 1975).

MOREAU, B., *Inventaire chronologique des éditions parisiennes du XVI*[e] *siècle* (Paris, 1972–).

MORTIMER, R., *Harvard College Library Dept. of Printing and Graphic Arts Catalogue of Books and Manuscripts*, i: *French 16th Century Books* (Cambridge, Mass., 1964).

MOSS, A., *Poetry and Fable: Studies in Mythological Narrative in Sixteenth-Century France* (Cambridge, 1984).

—— 'Fabulous Narrations in the *Concorde des deux langages* of Jean Lemaire de Belges', in N. Kenny (ed.), *Philosophical Fictions and the French Renaissance* (London, 1991), 17–28.

Muhlethaler, J.-C., *Poétiques du quinzième siècle: Situation de François Villon et Michault Taillevent* (Paris, 1983).

Munn, K. M., *A Contribution to the Study of Jean Lemaire de Belges: A Critical Study of Bio-bibliographical Data, Including a Transcript of Various Unpublished Works* (Scottdale, Pa., 1936).

Nève de Roden, A. C. de, 'Les *Mémoires* de Jean de Haynin: Des "mémoires", un *livre*', in C. Thiry (ed.), *'A l'heure encore de mon escrire': Aspects de la littérature de Bourgogne sous Philippe le Bon et Charles le Téméraire*, Les Lettres romanes, hors série (Louvain-la-Neuve, 1997), 31–52.

Nichols, S. G., 'The New Philology: Introduction: Philology in a Manuscript Culture', *Speculum*, 65 (1990), 1–10.

Nijhoff, W., and Kronenberg, M. E., *Nederlandsche Bibliographie van 1500 tot 1540* (The Hague, 1919–61).

Nitzsche, J. C., *The Genius Figure in Antiquity and the Middle Ages* (New York, 1975).

Norrell, R., 'L'Énigme de *La Concorde des deux langages*', *Revue des langues romanes*, 83 (1978), 151–5.

Omont, H., *Bibliothèque Nationale. Catalogue général des manuscrits français: Nouvelles Acquisitions françaises* (Paris, 1899–1918).

Ong, W. J., *Interfaces of the Word: Studies in the Evolution of Consciousness and Culture* (Ithaca, NY, 1977).

—— *Orality and Literacy: The Technologizing of the Word* (London, 1982).

Pächt, O., and Thoss, D., *Die illuminierten Handschriften und Inkunabeln der Österreichischen Nationalbibliothek: Französische Schule II* (Vienna, 1977).

Parent, A., *Les Métiers du livre à Paris au XVIe siècle (1535–1560)* (Geneva, 1974).

Parkes, M. B., *Scribes, Scripts and Readers: Studies in the Communication, Presentation and Dissemination of Medieval Texts* (London, 1991).

—— *Pause and Effect: An Introduction to the History of Punctuation in the West* (Aldershot, 1992).

Patch, H. R., *The Goddess Fortuna in Mediaeval Literature* (Cambridge, Mass., 1927).

Paxson, J. J., *The Poetics of Personification* (Cambridge, 1994).

Pensom, R., 'Zumthor and After: A Survey of Some Current Trends in the Reading of Old French Literature', *Medium Aevum*, 64 (1993), 294–306.

Picot, É., and Piaget, A., 'Une supercherie d'Antoine Vérard: *Les Regnars traversans* de Jehan Bouchet', *Romania*, 22 (1893), 244–60.

Poirion, D., *Le Poète et le prince: L'Évolution du lyrisme courtois de Guillaume de Machaut à Charles d'Orléans* (Paris, 1965).

Prince, G., *A Dictionary of Narratology* (Lincoln, Nebr., 1987).

Randall, M., *Building Resemblance: Analogical Imagery in the Early French Renaissance* (Baltimore, 1996).

Rawles, S. P. J., 'Denis Janot, Parisian Printer and Bookseller (*fl.* 1529–1544): A Bibliographical Study', Ph.D. thesis (Warwick, 1976).

Raynaud de Lage, G., '*Natura* et *Genius*, chez Jean de Meung et chez Jean Lemaire de Belges', *MA* 58 (1952), 125–43.

RÉAU, L., *Iconographie de l'art chrétien* (Paris, 1955–9).

REGALADO, N. F., 'La Fonction poétique des noms propres dans le *Testament* de François Villon', *CAIEF* 32 (1980), 51–68.

—— '*Effet de réel, effet du réel*: Representation and Reference in Villon's *Testament*', *Yale French Studies*, 70 (1986), 63–77.

—— '"En l'an de mon trentiesme aage": Date, Deixis, and Moral Vision in Villon's *Testament*', in E. Baumgartner, G. Di Stefano, F. Ferrand, S. Lusignan, C. Marchello-Nizia, and M. Perret (eds.), *Le Nombre du temps: En hommage à Paul Zumthor* (Paris, 1988), 237–46.

—— 'Gathering the Works: The *Œuvres de Villon* and the Intergeneric Passage of the Medieval French Lyric into Single-Author Collections', *L'Esprit créateur*, 33 (1993), 87–100.

RICHARD, A., 'Notes biographiques sur les Bouchet, imprimeurs et procureurs à Poitiers au XVIe siècle', *Bulletin de la Société des Antiquaires de l'Ouest*, ser. 3, 2 (1910–12), 544–61.

RIGOLOT, F., 'Jean Lemaire de Belges: Concorde ou discorde des deux langages?', *JMRS* 3 (1973), 165–75.

—— *Poétique et onomastique: L'Exemple de la Renaissance* (Geneva, 1977).

—— *Le Texte de la Renaissance: Des rhétoriqueurs à Montaigne* (Geneva, 1982).

ROSE, M. A., *Parody: Ancient, Modern, and Post-modern* (Cambridge, 1993).

ROTHSTEIN, M., 'Jean Lemaire de Belges' *Illustrations de Gaule et singularitez de Troyes* [*sic*]: Politics and Unity', *BHR* 52 (1990), 593–609.

RUPPEN, A., 'Bacchus René: Du "livre" au "moi" dans quelques préfaces de traducteurs à la Renaissance', *Versants*, 15 (1989), 21–33.

RUSSELL, D., *Emblematic Structures in Renaissance French Culture* (Toronto, 1995).

SAENGER, P., *Space between Words: The Origins of Silent Reading* (Stanford, Calif., 1997).

—— and HEINLEN, M., 'Incunable Description and its Implication for the Analysis of Fifteenth-Century Reading Habits', in S. L. Hindman (ed.), *Printing the Written Word* (Ithaca, NY, 1991), 225–58.

SANSY, D., 'Texte et image dans les incunables français', *Médiévales*, 22–3 (1992), 47–70.

SCHAPIRO, M., *Words and Pictures: On the Literal and the Symbolic in the Illustration of a Text* (The Hague, 1973).

SCHEIDEGGER, J. R., 'La Lettre du nom: L'Anthroponymie de Jean Molinet', *MF* 8–9 (1981), 198–235.

SCHELLER, R. W., 'Imperial Themes in Art and Literature of the Early French Renaissance: The Period of Charles VIII', *Simiolus*, 12 (1981–2), 5–69.

—— 'Ensigns of Authority: French Royal Symbolism in the Age of Louis XII', *Simiolus*, 13 (1983), 75–141.

SCHUTZ, A. H., *Vernacular Books in Parisian Private Libraries of the 16th Century According to the Notarial Inventories* (Chapel Hill, NC, 1955).

SCOLLEN-JIMACK, C., 'Marot and Deschamps: The Rhetoric of Misfortune', *French Studies*, 42 (1988), 21–32.

SEGUIN, J.-P., *L'Information en France, de Louis XII à Henri II* (Geneva, 1961).

SERVANT, H., 'Jehan de Liège, premier imprimeur valenciennois', *Valentiana*, 1 (June 1988), 6–11.

——'Un poème inédit de Molinet à la Bibliothèque Municipale de Valenciennes', *Valentiana*, 12 (December 1993), 3–8.

SEZNEC, J., *La Survivance des dieux antiques: Essai sur le rôle de la tradition mythologique dans l'humanisme et dans l'art de la Renaissance* (London, 1940).

SMALL, G., 'Chroniqueurs et culture historique au bas Moyen Âge', in L. Nys and A. Salamagne (eds.), *Valenciennes aux XIVe et XVe siècles: Art et histoire* (Valenciennes, 1996), 271–96.

Sotheby's (auctioneers, London), *The Collection of Otto Schäfer Part II: Parisian Books, Sold by Order of the Dr. Otto-Schäfer-Stiftung e.V., London, Tuesday, June 27, 1995* (Abingdon, 1995).

STEPHENS, W., *Giants in Those Days: Folklore, Ancient History, and Nationalism* (Lincoln, Nebr., 1989).

STEWART, S., *On Longing: Narratives of the Miniature, the Gigantic, the Souvenir, the Collection* (Baltimore, 1984).

STONE, D., 'Some Observations on the Text and Possible Meanings of Lemaire de Belges' *La Concorde des deux langages*', *BHR* 55 (1993), 65–76.

SULEIMAN, S. R., 'Introduction: Varieties of Audience-Oriented Criticism', in S. R. Suleiman and I. Crosman (eds.), *The Reader in the Text: Essays on Audience and Interpretation* (Princeton, 1980), 3–45.

SUSONG, G., 'Les Impressions arthuriennes françaises (1488–1591) et la grande rhétorique', *MF* 34 (1994), 189–203.

SUTCH, S. S., 'Comment lire une allégorie de personnification séculière', in *Les Grands Rhétoriqueurs*, i (Milan, 1985), 49–59.

TAYLOR, J. H. M., 'Un miroer salutaire', in J. H. M. Taylor (ed.), *Dies illa: Death in the Middle Ages. Proceedings of the 1983 Manchester Colloquium* (Liverpool, 1984), 29–43.

TCHEMERZINE, A., *Bibliographie d'éditions originales et rares d'auteurs français des XVe, XVIe, XVIIe et XVIIIe siècles* (2nd edn. Paris, 1977).

TERVARENT, G. DE, *Attributs et symboles dans l'art profane: Dictionnaire d'un langage perdu (1450–1600)* (2nd edn. Geneva, 1997).

THIRY, C., 'L'Honneur et l'Empire: A propos des poèmes de langue française sur la bataille de Pavie', in *Mélanges à la mémoire de Franco Simone*, i: *France et Italie dans la culture européenne: Moyen Âge et Renaissance* (Geneva, 1980), 297–324.

——'La Poétique des grands rhétoriqueurs', *MA* 86 (1980), 117–33.

——'Le Vieux Renard et le jeune loup: L'Évolution interne de la *Recollection des merveilleuses advenues*', *MA* 90 (1984), 455–85.

——'Rhétoriqueurs de Bourgogne, rhétoriqueurs de France: Convergences, divergences?', in J. Koopmans, M. A. Meadow, K. Meerhoff, and M. Spies (eds.), *Rhetoric—Rhétoriqueurs—Rederijkers* (Amsterdam, 1995), 101–16.

TOBLER, A., and LOMMATZSCH, E., *Altfranzösisches Wörterbuch* (Berlin and Wiesbaden, 1915–).

TOURNOY-THOEN, G., 'Deux épîtres inédites de Fausto Andrelini et l'auteur du *Iulius Exclusus*', *Humanistica Lovaniensa*, 18 (1969), 43–75.

——'Fausto Andrelini et la cour de France', in *Colloque international de Tours (XIVe stage): L'Humanisme français au début de la Renaissance* (Paris, 1973), 65–79.

WADSWORTH, J. B., 'Jean Lemaire de Belges and the Death of Pierre de Bourbon', *RN* 1 (1959), 53–8.

——*Lyons 1473–1503: The Beginnings of Cosmopolitanism* (Cambridge, Mass., 1962).

WARNING, R., *Rezeptionsästhetik* (Munich, 1975).

WAUGH, P., *Metafiction: The Theory and Practice of Self-Conscious Fiction* (London, 1984).

WEINBERG, B., *Critical Prefaces of the French Renaissance* (Evanston, Ill., 1950).

WIDMANN, H., *Vom Nutzen und Nachteil der Erfindung des Buchdrucks—aus der Sicht der Zeitgenossen des Erfinders* (Mainz, 1973).

WILLIAMS, A., *Clément Marot: Figure, Text and Intertext* (Lampeter, 1990).

WINN, M. B., 'In Pursuit of the *Acteur* in *La Chasse et le départ d'Amours*', *MF* 7 (1980), 92–117.

——'Publisher vs. Author: Anthoine Vérard, Jean Bouchet, and *L'Amoureux transy*', *BHR* 50 (1988), 39–55.

——*Anthoine Vérard, Parisian Publisher, 1485–1512: Prologues, Poems, and Presentations* (Geneva, 1997).

WIRTH, J., *L'Image médiévale: Naissance et développements (VIe–XVe siècle)* (Paris, 1989).

WOLLEN, P., *Raiding the Icebox: Reflections on Twentieth-Century Culture* (London, 1993).

YATES, F. A., *The Art of Memory* (London, 1966).

ZSUPPÁN, C. M., 'An Early Example of the Renaissance Themes of Immortality and Divine Inspiration: The Work of Jean Robertet', *BHR* 28 (1966), 553–63.

ZUMTHOR, P., *Le Masque et la lumière: La Poétique des grands rhétoriqueurs* (Paris, 1978).

INDEX

Note: *Italic* numbers denote reference to illustrations.

Printed in the USA/Agawam, MA
October 1, 2019

739552.024